THE
BREAST
ADVICE

THE BREAST ADVICE

ALL YOU NEED TO KNOW ABOUT BREAST HEALTH, SCREENING, AND TREATMENT

DR. ELISA PORT

HarperOne

An Imprint of HarperCollins*Publishers*

TO MY HUSBAND: We share the calling and passion for curing cancer and for trying to make the world a better place for our patients and beyond. Thank you for your love and support always and in all ways.

TO MY CHILDREN: You grew up sharing your parents with their patients, and yet somehow turned out to be smart, nice, and seemingly well-adjusted. Lauren, you are following in our footsteps to share the beautiful field of medicine. And Zachary, you contribute in so many ways that make us proud. My hope is that you both find the same meaning and fulfillment from your pursuits that your parents have.

TO MY PARENTS: Jeff and Loni Rush, for your love and support from the beginning. Early in my life, you guided me back to medicine when I was considering a career in making music videos. It definitely worked out better this way.

AND OF COURSE, TO MY PATIENTS: Both those who shared their stories here, and those who didn't. You have *all* given me the honor of taking care of you. You have all taught me so much, and you have all contributed to this book in some way.

CONTENTS

Introduction   *1*

PART I: BREAST HEALTH

WHAT IS BREAST HEALTH?   *13*

SECTION 1: LIFESTYLE   *19*

1   Was It the Sugar?   *21*

2   How Are Hormones Involved: Are They a Godsend or Should You Proceed with Caution?   *37*

3   Supplements: Is There a Secret Potion to Give You the Edge?   *48*

4   Stress: My Divorce Gave Me Breast Cancer, and Other Myths   *57*

The Breast Lifestyle Advice   *63*

SECTION 2: SCREENING   *65*

5   Mammograms: How to Make Sense of the Moving Target That Is the Guidelines for Screening   *67*

6   More Screening Options: Where Do Ultrasounds or Sonograms Fit In?   *79*

7   MRIs: For Screening High-Risk Patients, When Diagnosed with Breast Cancer, and for Follow-Up   *85*

8   A Bit About Biopsies, and Some of the *Normal* Things We Might Find   *93*

9   Cutting-Edge Screening Technology: Full-Body Scans, Liquid Biopsies, Thermography—What's Ready for Prime Time and What Isn't?   *100*

10   If You Feel Something, Say Something!   *110*

The Breast Screening Advice   *120*

SECTION 3: HIGH RISK *123*

11 Genetic Testing: Knowledge Is Power *125*

12 When It Comes to Genetics, Don't Forget Your Father's Side *135*

13 Other Risk Factors for Breast Cancer: Family History, Atypia, LCIS, and Breast Density *143*

14 Breast Cancer Risk Assessment Tools: Screening and Prevention Strategies for Those Who Are at Increased Risk *153*

The Breast High-Risk Advice *162*

PART II: BREAST CANCER

SECTION 4: THE BASICS OF BREAST CANCER *167*

15 What Is Breast Cancer? *169*

16 How to Be a "Breast Friend" to Someone Who Has Been Diagnosed *178*

SECTION 5: SURGERY *185*

17 Lumpectomy vs. Mastectomy and Making the Big Decision *187*

18 One Breast or Both? *200*

19 Save the Nipple?: Choosing Which Type of Mastectomy Makes Sense for You *210*

20 To Reconstruct or Not to Reconstruct? *216*

21 Lymph Node Surgery: Is It Always Necessary and, If So, How Much? *229*

22 Treating DCIS: Understanding the Earliest Form of Breast Cancer *240*

23 Is Breast Cancer Overtreatment a Thing?: When Might You Choose the "Watch and Wait" Approach? *247*

The Breast Surgery Advice *251*

SECTION 6: TREATMENTS BEYOND SURGERY *255*

24 Chemotherapy: Longer and Stronger vs. Dialing It Down *257*

25 Chemotherapy First or Surgery First? *273*

26 Radiation *281*

27 Antihormonal Treatment *295*

28 All the Latest Novel Therapies *305*

29 Breast Cancer Follow-Up and Can It Come Back? *313*

THE BREAST TREATMENT ADVICE *321*

SECTION 7: SPECIAL SITUATIONS *327*

30 Rare Types of Breast Cancer *329*

31 Young Women with Breast Cancer: How Is It Different? *335*

32 Fertility Preservation: Keeping All the Options Open *342*

33 Disparities in Breast Cancer Care (and Health Care in General) *347*

34 Male Breast Cancer *356*

35 Talking to Our Children, Friends, and Colleagues: What Do They Need to Know About Your Breast Cancer? *366*

SILVER LININGS: BREAST CANCER CAN CREATE A TEACHABLE MOMENT *381*

Acknowledgments *391*

References *394*

Index *398*

INTRODUCTION

When I was first starting out in practice, with a few years as a breast cancer surgeon under my belt, a patient named Susan came into my office, accompanied by her husband and two sisters. She was about fifty years old with multiple spots of cancer in one breast, needing her whole breast removed: a mastectomy. As we talked about the surgery, the recovery and the process, Susan seemed ready and more important, very optimistic. "I know I can get through this!" she told me, surrounded by her family. "Let's get it done."

I had come out of medical school and training extremely well prepared, I felt, to take excellent care of patients. I had done countless breast operations and learned from the best, starting my practice at one of the top cancer hospitals in the country. The care of breast cancer patients as my specialty was something I arrived at after four years of medical school, five years of surgery residency, two years of research in a breast cancer laboratory, and a year of intense training in breast surgery. I knew this specialty was for me for every reason: the technical operative skills I had honed, the privilege of developing relationships with patients at one of the most critical junctures of their lives, and becoming part of a field where there was so much room for optimism. There seemed to be new options for treatments and progress being made every year. I knew from the beginning that I wanted to be part of all that, and when it came right down to it, take care of women and cure their cancers. Or as my young children used to explain to others on the frequent occasions I was not around, "My mommy is taking care of sick mommies so they can go home to their children."

But as it turned out, I still had something to learn: specifically from my interactions with my patients. The week after her first visit to my office, Susan had her surgery, and all went really well. She was up and around as expected even the day of surgery: walking, talking, eating, drinking, going to the bathroom. When I visited her the morning after her surgery, she was thrilled to have this hurdle behind her and ready to go home. While she had some pain, it was manageable, and like many patients, she had a sense of euphoria from having cleared this major obstacle, surgery, and also a bit related to residual anesthesia.

Susan came back a week after surgery, as I have all my patients do, to go over her results from surgery, and given her excellent report, she beamed. She confirmed that from a recovery standpoint, she continued to do really well, doing more and feeling better every day.

Three months later, she returned for a follow-up visit. This time, she seemed a little more dejected than right after surgery. "You know, I went home after surgery and did great," she informed me. "But I gotta tell you, as I started to recover more and more, and do more and more, I found myself feeling extremely tired. I would do stuff for a few hours, and then out of nowhere, this wave of fatigue would sweep over me. I was pooped and needed to lie down! It felt like a setback. Maybe I went back to the office full time too early?" Then she went on to share with me, "I didn't realize and don't think I was fully prepared for how tiring recovering from surgery is. I'm still not one hundred percent."

I thought about this for a moment and understood why this was true. "You know, Susan, your body experiences surgery as a trauma, as an injury. Repairing wounds, either unintentional—like in a big accident—or intentional like surgery to cure cancer, involves your body diverting a lot of energy sources to healing. We know that from our trauma patients. Everything we do consumes energy. Just being awake consumes energy! And sleep can be one of the most reparative, restorative things you can do while recovering, so all the energy stores you have can be directed to healing, rather than diverting them to the many activities that we do when we are awake. So, when your body needs to heal, it sends the message to power down and rest. That's why you are tired. Listen to your body telling you what it needs to do to heal."

I could tell Susan was listening, and that what I was saying was registering, bringing her experience together for her. "That makes total sense to me," she said. "It's exhausting. Recovery from surgery is exhausting! You should always tell your patients that before you send them home after surgery!" And ever since then, I do exactly that. It was one of the first times that I can remember that a patient gave *me* meaningful advice and information—the kind you don't necessarily learn in medical school or surgical training—that I have since imparted to so many others.

Chances are, if you're reading this, you're either concerned about your risk of getting breast cancer or have been recently diagnosed. Or perhaps it's your friend or family member going through it. I wrote this book for you, but I didn't write it alone. In the pages to follow, you're going to find up-to-date, highly informed medical advice about breast health and breast cancer, but that's not the only expertise featured here. You're also going to hear from *my patients*, in their own words, speaking from the heart about their breast cancer journeys, because although I can bring the doctor's perspective, I've never had to deal with a breast cancer diagnosis myself. And as you're about to discover, there's so much to learn from patients, like Susan, who've been there.

Breast cancer is the most common cancer affecting women in the United States. There are 300,000 women diagnosed in the United States each year, and of *all* cancers women get, one in three will be breast cancer. Given that statistic, everyone has been touched by breast cancer in some way. If you haven't been personally affected, you probably know someone or know someone who knows someone who has.

In the time I've been in practice, I'm happy to say I've watched as the field has moved at warp speed. The overall survival rate of breast cancer is now upward of 90 percent. In some cases, we can give women a 99 percent chance of survival, excellent news and absolutely a source of optimism if ever there was one in the cancer world! Every day in my office, I'm able to tell my patients that there's an incredibly high chance that they will not only survive but thrive after a breast cancer diagnosis.

And every day, my patients tell me stories and give me input, advice, or insights just as Susan did. From one patient, I learned to tell women to bring a button-up shirt to the hospital, so they don't have to try to pull a T-shirt over their heads with a limited range of motion after surgery. Another gave me crucial insight into navigating "life" during chemotherapy. She told me, "I didn't want to become a recluse during treatment, even though I was super exhausted. In my life before chemo, I would meet friends for drinks, then dinner, then out dancing. During chemo, it was just dinner. This allowed me to stay connected to my friends and keep up some semblance of a social life without overdoing it. This was my chemo-hack." Another patient taught me about "joy mining," digging deep for a nugget of happiness or a source of joy, no matter how small, even on the darkest, toughest days of treatment. My patients tell me that they wouldn't have gotten through it with their mental health at least somewhat intact without hearing the stories and advice of their fellow patients and survivors, who have walked in their shoes and understand the toll all forms of treatment can take.

There are also over four million breast cancer survivors living in America who continue to think about their diagnosis, and it affects women profoundly long after they have been treated and cured. As a result, almost every patient I've ever met has benefited in some way from the support of the breast cancer community, whether online or in real life. As you'll learn in the book, I have two patients who met in my waiting room who have remained friends ever since and schedule their annual follow-up appointments together so they can go and get lunch after. This disease that predominantly affects women—and has so many survivors—has generated a community that is legendary in its strength. Someone who is newly diagnosed with either a genetic mutation or with breast cancer has a community at their fingertips and can immediately draw on it for comfort, input, and support. They can find people who will help them feel less alone and tell them everything is going to be okay in the end; where hacks and other advice are shared from patient to patient.

The only problem is that while patients can reassure one another, they can also make one another more anxious by sharing treatment information that might not be applicable from one patient to the next. That's because breast cancer is not a one-size-fits-all disease, and no two cases are precisely the same. So what works for one patient doesn't necessarily translate to another one. Only recently, a patient scheduled for a lumpectomy, Jess, shared with me a story on Instagram of another woman who seemed to have "the exact same thing" opting for a double mastectomy. In the comments, everyone was cheering on the woman who had chosen the mastectomy, and so my patient showed the post to me wondering if she should be having a double mastectomy as well. I reminded Jess that the outcomes for her type of breast cancer with a lumpectomy are excellent and that a mastectomy was going to be much more invasive with a longer recovery time—but that everyone has a right to make their own medical decisions in consultation with their doctor, and the woman on Instagram had likely chosen what she thought was best for *her*. It was a reminder that patients can have strong effects on one another, but not all of them will be positive. Sometimes, hearing about other women who have taken another treatment path can plant seeds of doubt and confusion in your mind, impeding you from proceeding with your own course of treatment with confidence, which is exactly what had happened to Jess.

The proliferation of advice online and in person can also make it hard to separate fact from fiction, not only about breast cancer in general, but more importantly, from case to case. And when misinformation circulates, especially online, it can quickly spread. Time and again, I've seen inaccurate information doing the rounds without anyone putting their hand up and saying, "Hey, that's just not true." Misinformation abounds about the best tests to screen for breast cancer, treatments for breast cancer and their side effects, and the benefits of different surgical approaches, to name just a few.

I remember one patient, Sam, who first came to see me at thirty years old after genetic testing determined that she was at the highest risk for developing breast cancer, testing positive for the BRCA1 gene. Her risk of getting breast cancer was up to 80 percent, and she was also at high risk for developing it at a young age. As a result, she had decided to move ahead with a preventive double mastectomy, entirely appropriate in her situation. Sam had just gotten married to an extremely supportive spouse. Together, they were wrapping their heads around the idea of the operation, its magnitude, and optimal timing. They were also planning a delayed honeymoon to Italy and had decided to schedule the surgery for the week after they got back.

"I'm ready," Sam said to me definitively, her new husband nodding his head in agreement. "Doing the surgery right after I get back from Italy is a bit of a drag, given that I can't have a drink for the three weeks leading up to the surgery date, but I'll manage. It just would have been nice to have a couple of glasses of wine on vacay. Or a negroni, my favorite cocktail." I paused for a moment before responding. Had I heard correctly?

"Wait, what? What did you say about three weeks and no drinking? Where did you hear that?" I asked.

"Well, I have been spending some time in this group chat for BRCA-positive women who are gearing up for their surgery, and a few who have already had it and are trying to support the others. It's a wonderful source of information. In fact, I met one of your other patients through this platform. We connected in person, and she showed me her results, and they look amazing. She's been there for me ever since, and it's really helped me to feel like I'm not alone in the journey. She's part of the reason I felt comfortable and ready to move ahead."

"But go back to that thing about the drinking . . ." I said.

"Oh yeah, so this third person and someone in our group chat said her surgeon told her it's dangerous to drink any alcohol in the three weeks leading up to surgery, and she shared that with us. Your team didn't tell me that."

"That's because it's not true," I answered.

"So, why would her surgeon tell her that?"

"I'm not sure, Sam. People have a lot of interesting or different practices; maybe her doctor had some particular concern about her case? However, drinking in moderation leading up to the surgery, in general, does not affect outcome. *After* surgery you wouldn't want to mix alcohol and pain medication; do you think that's what she might have meant? After?"

"No, definitely not. Wow."

I explained that this is an example of misinformation, of people taking a small shred of information, like the fact that the risk of breast cancer increases with alcohol intake, and letting it morph into something else. There are no data to suggest that having a few drinks *leading up to surgery* is harmful. I certainly didn't understand where the arbitrary "three weeks" number came from. Why not four weeks? Why not two? Because none of it is true.

Thankfully, the advice Sam got about abstaining from alcohol never would have harmed her health, but other advice she might have gotten, such as forgoing a mammogram to instead do some other unproven test, could have had a much more negative impact. So yes, while patients can support each other, they can also freak each other out. There are millions of posts, groups, podcasts, and articles out there about breast cancer, making it even harder for patients to get the support and comfort they need while also accessing medically informed and up-to-date advice.

This is where this book, *The Breast Advice*, comes in. In every chapter of this book, you're going to find stories of patients talking about what they've learned and sharing pearls of wisdom, *and* you're going to find accurate and up-to-date, trusted medical advice and insight on each breast cancer–related topic from me, a surgeon with more than twenty-five years in the field. By bringing patient wisdom and doctor expertise together, my goal is to give you the most comprehensive and reassuring guide to breast cancer risk, screening, diagnosis, and treatment available.

Part one of this book focuses on breast health. Here, my patients and I will be talking about prevention, screening, and lifestyle. Perhaps more than any other cancer, we know exactly how to "get ahead" of breast cancer, thanks to regular screenings and paying attention to *certain* lifestyle factors such as diet, exercise, alcohol consumption, and hormones. You're going to learn that when it comes to lifestyle, it's all about balance. As it turns out, it's not that mysterious or tricky: breast health is total body health—in other words, what's good for your breasts is good for the rest of you too. In this section of the book, we're also going to talk about what constitutes high risk, and the situations where early screening and/or genetic testing may be necessary, cutting through the noise and misinformation with advice rooted in data, science, and hard-won experience including up-to-the-minute developments, like incorporating AI into screening for breast cancer.

In part two of the book, my patients and I focus on advice for those who have had a recent diagnosis. Here's where we will dig into breast cancer treatment and management. My patients will talk movingly about their journey through breast cancer, and

I'll speak to the many choices that women who are undergoing treatment have to make: Lumpectomy versus mastectomy? Single or double mastectomy? Chemotherapy first or surgery first? So many patients ask me what I would do myself if I were in their situation. I always say, "I may be the expert in cancer, but you are the expert in *you*." And that we will chart their course together, knowing that I will only offer them safe options. In the course of this book, we're going to cover all these topics and more, along with information about the newest treatments.

Later in the book, we will cover special situations such as why breast cancer is increasingly common in young women and why Black women are at higher risk for more aggressive breast cancers. I also discuss disparities in breast health care and the rare but important entity of male breast cancer. Each new chapter provides the latest medical information in conjunction with a patient story—sharing advice or experience—on each of these fronts. Since misinformation can be a common source of doom and gloom, we're also going to focus on information and progress, both of which give us huge reasons to be optimistic, what I call informed optimism.

So much of what I do on a one-to-one basis for patients is trying to connect what people have learned and heard outside my office with what is or is not true. I provide them with the facts and reliable information that empower the best decision-making. Of all the messages I would hope you take away from reading this book, here are the two big ones. You have every reason to be optimistic—and you don't have to go it alone. There are so many people out there who have either walked this path or are currently walking this path. Find your "breast friend" and in the meantime, you have this book to be your trusted guide. Come along with us—we can get through this, together.

BREAST HEALTH

What Is Breast Health?

We have all heard of heart health and mental health, but what is breast health? To put it simply, breast health involves learning about and understanding what does and does not affect your risk of getting breast cancer, and doing what you can to prevent it or reduce your risk of getting it. There is so much information out there! In the first part of the book, we're going to discuss the factors that can affect the risk of getting breast cancer, and talk about what the science actually shows related to breast cancer risk. The second part is personalizing what does and does not ring true for *you* as it relates to these risk factors. As you will soon learn, there are factors that affect our risk of getting breast cancer that we can control and influence, and others that we can't. Understanding the actionable ways that breast cancer risk can be reduced is empowering. But even for things you can't control, like your family history, it's also empowering just to understand what these factors mean for you. Knowledge *is* power and can lead to action. Breast health also involves knowing what to look for. Some of the warning signs of breast cancer are more obvious than others, but they are all important to be aware of and call attention to if identified:

- **A mass or lump:** If you notice a mass or a lump in your breast, this requires attention. By lump or bump I mean a more prominent area that might feel as though it is protruding from the surrounding tissue in your breast; something that you notice because it sticks out against your normal background. This is why it's important to get familiar with your breasts, checking yourself once a month or so, in order to know your body and be familiar with what feels normal for you. While many lumps are not cancer and can be other benign (noncancerous) things like cysts or benign tumors, no lump should be ignored or dismissed without further investigation.

- **Skin dimpling or changes:** Sometimes a mass doesn't feel so obvious, but you might notice indentation or dimpling on the skin's surface. This can be a cancer underneath, pulling and tethering the overlying skin. So, pay attention to the appearance of your breasts, not only how they feel.

- **Nipple discharge:** Many women experience forms of nipple discharge, especially in and around periods of breastfeeding. If you are not breastfeeding, new nipple discharge is concerning, and in a small percentage of cases can be related to a new cancer and therefore needs to be further investigated. Here are three red flags that constitute concerning nipple discharge: (1) it's only on one side; (2) it's bloody; and (3) it comes out spontaneously, without squeezing the breast. A stain in your bra or on your pajamas can be a sign of spontaneous discharge. Any new nipple discharge should be brought to a doctor's attention, even if only one of these criteria holds true.

- **An enlarged lymph node under the arm:** For a small number of cases, breast cancer can first appear only after it has spread to a lymph node in the armpit. In such cases, it's possible you notice a lump under the arm even before there appears to be anything abnormal or changed in the breast.

- **Changes in the breast skin or the nipple:** Rarer forms of breast cancer can first show up as skin changes—such as redness or a rash—or changes to the nipple like crusting or scabbing. In addition, while many women have one or both inverted nipples as their norm, if your nipple becomes inverted—when it had not been previously—this can be a sign of a cancer developing behind it and pulling the nipple inward. While new findings such as these can be normal changes, and a rash can be just a rash from leaving your sweaty workout bra on too long, for example, they can also be the first signs of breast cancer.

Any of the signs above can be associated with normal changes to the breast or anatomy, but they are important to get checked out to optimize breast health and maximize chances for early detection if cancer does develop.

Finally, breast health involves screening. One way to keep yourself as healthy as possible is to make sure you are up-to-date with all of your imaging. While there are many health issues and types of cancer for which there is no effective screening or early detection, breast cancer is not one of them! In the ensuing chapters you will learn about why and how screening works, and that while the general guidelines are to start mammograms at age forty—and to have them yearly after that—there is no one-size-fits-all. Adding on additional tests might be important for you based on many personal health factors such as your family history, your breast density, and yes, your overall health. You will learn all about whether these additional tests, sonograms and/or MRIs, contrast-enhanced mammograms, and even newer AI technology, should be considered in your particular case, and how you can empower yourself to take the lead and ask your own doctor whether doing additional imaging is appropriate for you.

Keeping you "breast healthy" can involve finding new things or abnormalities on these imaging studies. Since early detection is the key to improved survival, it's important to know that as with every other aspect of your health, and other screening practices, if something concerning is found, often more investigation is needed: added tests, follow-ups, or even a biopsy. These are all very common and routine parts of breast health practice to rule out a new cancer. It's relieving to know that if you are called back for additional tests after a mammogram, 90 percent of the time the additional tests will show there is nothing of concern, and the issue will be resolved. Even if a biopsy is needed, the majority of these, up to 80 percent, will prove to be benign, not cancer.

Many of my patients, especially those who have had breast cancer, come in every year incredibly anxious about having their mammogram or whatever additional tests they are scheduled for. Some tell me they don't sleep well the night before their yearly visit with me, which is often coupled with screening tests. When my patients receive a normal report, the first word I blurt out when I walk through their exam room door is "Congratulations!" because I know that is what they are waiting to hear. Their relief is often palpable, manifesting as a deep exhale, a big smile, or even happy tears. I never get tired of giving good news. Accepting these results—good or bad—and taking follow-up steps, are *all* part of being on top of your breast health. Learning how to navigate the ins and outs of lifestyle factors, personal history and health factors, and screening practices that do or do not apply to you will put you in the best position to optimize your breast health. Let's get to it.

LIFESTYLE

WHAT WE KNOW NOW ABOUT WHAT
DOES AND DOES NOT INCREASE
RISK OF BREAST CANCER

1

WAS IT THE SUGAR?

STACEY GRIFFITH, founding senior master instructor of SoulCycle: For me, being diagnosed with cancer was a trip. I could compare it with maybe a race car driver who gets in a car accident. You know, you're a professional driver, you drive fast for a living, you're always in control of the wheel, you never miss a turn, you know the direction. It was just paralyzing. I did not anticipate that. I did not expect that to come. It stopped me in my tracks, and I'd never had to deal with anything like that because the only thing I'd dealt with health-wise before were frozen shoulders, ACL tears, double Achilles tears, meniscus issues caused by my work. My cancer was not caused by anything I did, I don't think, unless there's something in those candles that we burn during SoulCycle class, who knows! Since this was the first major physical hiccup I had, I didn't really know how to handle it. My mental health became so compromised. It was hard for me to even form a sentence for a week, and I talk and coach for a living!

My employer could not have been more encouraging and accepting, giving me months off work. I hadn't *not* worked since I was fourteen years old. I don't know not working. I don't know sitting still. I don't know resting. I don't know hospitals. I don't know doctor visits. So, I had to make a shift, and I had to become a hospital goer. The hospital became my gym. I just kind of made it my gym. I made it my workout, and I knew I would get back to the other type of gym when I was ready.

CINDY M., environmental lawyer: When I was diagnosed with breast cancer with no family history, I started googling all the possible risk factors trying to understand how this had happened to me. None of the things I read about related to my history applied to me: I didn't get my period at an early age; I breastfed all my kids—there was nothing I could point to. So, naturally, I assumed it must have been something I did or had been doing related to my lifestyle. Maybe I was eating too much sugar? Not enough vegetables? Too much red meat? Too much soy? Not enough soy? There had to be *something* I did, a smoking gun, that contributed to my getting breast cancer! And therefore, something I could change to make sure the cancer didn't come back. So, I decided to eliminate processed sugar from my diet, stopped drinking completely, and started going to the gym every other day. I needed to feel like I was doing *something* to prevent breast cancer from coming back, some form of control over my outcome. There certainly is enough information out there to suggest that these changes could make a difference.

Dr. Port told me the data were mixed as to whether or not these lifestyle changes would have a strong effect or move the needle on my risk of recurrence, but for sure she agreed that all of these things contributed to a healthier lifestyle on a bunch of different fronts, so why not?

Then there were all these other things people were advocating for that I considered doing, like supplements, anticancer diets, and IV drips. I went to Las Vegas and they were literally in the hotel, next to the nail salon, offering IV drips with all of these different additives: vitamins, minerals, cleaning out of toxic metals; there was a whole menu of choices. The immune booster, the fatigue remedy, and of course, the anticancer formula. When I asked the guy at the desk what was in each of these, the answer he gave me was it was their own secret recipes that had been tested extensively in their own laboratory. Located where? He had no idea. For me, this is where my skeptical instincts kicked in. I didn't want some random mall person putting a needle in my arm and infusing God-knows-what into me. It just crossed some sort of imaginary line in my head that this could potentially do more harm than good. Plus, it was expensive; the price of one infusion equaled one night in my Vegas hotel.

So, I stuck with trying to do better with the stuff I *was* willing to put into my body. What I ate and drank or didn't drink. Even if I now know that for me, and most women, a breast cancer diagnosis cannot be traced back to something I did. It's not our fault.

DR. PORT: Consider this: almost 90 percent of women diagnosed with breast cancer have no family history or other identifiable risk factors when diagnosed. So, it's not surprising that patients often ask, "Why did this happen to me?" Many, like Cindy, go on to analyze every aspect of their lives in an attempt to get a handle on the *why*.

Did I grow up near a toxic waste dump? How come so many people in my neighborhood have been diagnosed? Is it from drinking water from plastic bottles or eating from Tupperware? What about my diet and how much I exercise?

My patients have asked me about all of these potential causes and so many more. Breast health—preventing breast cancer or reducing one's risk of getting it—is something that most women have thought about, some more than others, depending on their other risk factors. With millions of websites, podcasts, and social media posts out there, anyone can find fuel for *any* theory about what causes breast cancer, regardless of how scientific—or unscientific—it may be. Recommendations to "starve your cancer of sugar" or to drink only from glassware and avoid all plastic containers—so many others abound. As a doctor with a strong science background, *I* can often barely make sense of new information that's out there. Finding clear, accurate information is complicated, and there are so many conflicting reports. *Eat more soy. Eat less soy. Alcohol in moderation is safe and actually good for you. Alcohol is carcinogenic and it's best to abstain completely.* That's before we even talk about the supplements, immune boosters, and anticancer secret formulas hawked by nutritionists, snake oil salesmen, and even by doctors with respectable credentials alike (more on supplements later!). How is a person supposed to sift through this and find what's true on a normal day, let alone immediately after getting diagnosed with cancer?

The answer is to first take a deep breath. Then, look toward the data. Not just one shred of data, but what the majority of the data shows. One study on rats done in a lab does not change the way we practice medicine or provide conclusive evidence. Many sources of information or those trying to prove a point will cite one study. But in the medical field, rarely can we make sweeping or definitive conclusions about risk factors based on just one study. We look at the *preponderance* of data that shows to influence outcome repeatedly and reliably, in large numbers of patients from many studies.

Now that that's settled, let's focus on two things that are *known* to increase your risk of getting breast cancer and increase the risk of breast cancer coming back if you have already been diagnosed.

WEIGHT

Here's what a significant amount of data supports: being overweight or obese is the number one lifestyle factor that increases the risk of getting breast cancer, as well as the risk of breast cancer recurrence for those already diagnosed.

How are weight and breast cancer risk related? The short answer is it all circles back to estrogen: overweight people have greater fat stores, which produce more estrogen. Let me walk you through the specifics of the connection. Did you ever wonder why after menopause, when estrogen production and supply supposedly "shut off," we don't morph into men? Our ovaries no longer produce estrogen, *but* our voices don't deepen, we don't grow beards, and thankfully, we don't grow hair on our bodies the way men do. Think about it: With our ovaries (the main estrogen factory in our bodies) powered down, is there still estrogen circulating in our biologically female bodies? If so, where does it come from? The answer is, yes, there *is* still estrogen produced by our bodies, and one of the main sources is fat stores.

Fat stores are not just passive blobs of tissue that are attached to our bodies. They comprise metabolically active cells that produce and secrete lots of things, like hormones, that then get released into our bloodstreams, circulate around, and affect other tissues. Fat stores can also produce and secrete pro-inflammatory agents that have been implicated in all kinds of diseases: everything from cancer to heart disease, even Alzheimer's disease and dementia. Most importantly from a breast cancer standpoint, fat stores produce lots of estrogen. This estrogen derived from fat stores gets released into the bloodstream and circulates in the body, reaching *all parts* of the body, including our breast tissue and breast cells. Approximately 60 to 70 percent of all breast cancers are fed and fueled by hormones. So, increasing circulating estrogen definitely increases risk for developing breast cancer and causes estrogen-fed tumors to grow. This is the weight–breast cancer connection: fat stores produce estrogen, which feeds most breast tumors. It's the reason why maintaining a healthy body weight, without excessive fat stores, can both decrease the risk of getting breast cancer and decrease the risk of recurrence for those already diagnosed.

There are many definitions for a healthy body weight and of course body image, but there are certain boundaries where increased weight does cross over into being objectively unhealthy. Body mass index (BMI) is one way to assess weight, but there are others, including measuring waist circumference or waist-to-hip ratio. By any of these definitions, 30 to 40 percent of the adult American population are currently categorized as obese. Obesity increases health risks including heart disease, diabetes, stroke, and, yes, some cancers, including breast, colon, and uterine. These are factors to consider and discuss with your doctor.

Because maintaining a healthy body weight is so important,

there are numerous strategies offered up for how to do this. There are thousands of diets, fad and otherwise, and food pyramids promoting different ratios of food intake and types of foods to eat more of or less of. With all of the emphasis on components of a healthy diet, it is important to know that there is no *individual* dietary component, in and of itself, that can reduce the risk of getting breast cancer or recurrence. Let's talk about the superfoods: broccoli, blueberries, tofu, green juice. All are associated with a generally healthy diet, and all have been touted as having anticancer properties and playing important roles in an anticancer diet. These foods *are* associated with a healthy diet, but there is no magic dietary bullet that has been scientifically proven to reduce the risk of getting breast cancer or preventing it from recurring. Blueberries do have antioxidant properties and are rich in so many other elements that *may* reduce breast (and other cancer) risk. There have been studies of mice fed blueberry-rich diets where the animals have been found to develop fewer breast cancers. Like all basic science and laboratory studies, it's such a big leap to go from this to an actual preventive—much less curative—effect in human beings! At least for now, that leap cannot be made. But blueberries are delicious and certainly won't hurt you in moderate quantities.

If blueberries and broccoli have been misguidedly hyped as anticarcinogenic, sugar has been implicated as the opposite: a cancer promoter. Google "Does sugar feed breast cancer?" and you will get thousands of results in the affirmative. Like many myths, the idea that sugar feeds breast cancer originated from a shred of truth. It is true that cancer cells are typically more metabolically active than normal cells, growing, dividing, spreading, and doing their bad cancer things. There is even a test called an FDG-PET scan that can check for cancer spread by looking to see where labeled sugar molecules go. Tumors, which are more metabolically active, consuming higher amounts of sugar, do "light up" on these scans, and can give us information about cancer spread (more later on this type of scan and how it is used in practice). But all cells require sugar for fuel, and you cannot "starve" your cancer cells by depriving your body of sugar. Even on a sugar-free diet your body will convert other types of fuel to sugar to feed all of its cells, unfortunately including cancer cells. A *sugary* diet (filled with processed foods and desserts as well as sodas, juices, and other sweetened drinks for example) is often associated with higher weight, and therefore a potential contributor to breast cancer risk and recurrence, but only as it relates to weight, not sugar content per se. So, the idea of sugar feeding cancer is a myth that's spiraled out from the fact that maintaining a lower body weight reduces cancer risk. A piece of chocolate cake, a soda, or my personal favorite, a pack of M&M's, which I subsisted on during residency when I could buy them from the hospital vending machine to sustain me on my long overnight shifts, will not make or break one's overall health or breast health.

The same is true for a diet rich in fatty and processed foods. These types of dietary regimens often lead to being overweight and obese. So, the take-home message is most of the dietary factors that have been implicated in breast cancer risk all funnel into association with increased weight and are not necessarily related to a specific component of that diet that causes or cures breast cancer in and of itself.

EXERCISE

Exercise is definitely beneficial, and an essential component of an overall healthy lifestyle, both in terms of quality of life and potential quantity of life. Exercise is beneficial in part as it relates to weight control, but independently as well. Combining both aerobic and strength training is essential. Recently, I met with a patient in my office who was tall, thin, young, and newly diagnosed with breast cancer. When I started explaining to her that she couldn't exercise for a few days after her surgery, she said, "That's not a problem. I don't exercise. At all." That was a new one for me. Almost *all* women who maintain a healthy body weight do some form of exercise.

While it's tricky to pin the direct connection between exercise and breast cancer risk on one specific mechanism, research offers some insights. For example, exercise may lower the amount of circulating estrogen, which fuels some breast cancers, and may increase immune system activity. The benefits of regular exercise to overall health and almost every organ system are incalculable and well established: the heart, lungs, musculoskeletal system, and our mental health all benefit from regular exercise. These benefits are related to so many complex things that happen in our bodies and to our bodies when they are pushed. Without going deep into the biochemistry of it all, cortisol, glucose levels, endorphins, hormones, and dopamine and serotonin levels are all positively affected by almost all forms of exercise, toward making us healthier.

Not only does exercise make us healthier, it can make us *feel* better too. For example, many of the side effects from some of the medications we use to treat breast cancer, such as bone and joint pain, can actually be improved by exercise and movement. Going to the gym; hitting the weights; getting on the exercise bike, the treadmill, or the elliptical; taking fitness classes; jumping in the pool. These are all amazing examples of routines and activities that increase heart, lung, bone, mental health, and so much more. Importantly, there is no specific form of exercise that reduces cancer risk, and there is no magic exercise bullet here for preventing breast cancer. I have treated women who range from professional athletes to regular people who are serious gym devotees who have gotten breast cancer that their heavy exercise regimens did not prevent. Aerobic exercise, weight-bearing exercise, and strength training are all different aspects of exercise activity that are important, and most professional medical organizations currently recommend two and a half to five hours of exercise per week, with at least one to two hours being vigorous exercise, like running or swimming.

But let me put in a plug here for my personal favorite when at all possible: exercising outdoors. It's true the weather does not always cooperate, but "green exercise" is associated with even further health benefits and, not surprisingly, more likelihood of repeating the activity. Often, people engaging in outdoor activity are seeking an exceptional challenge: climbing a mountain, training at altitude, or some other extreme sport, but the benefits of engaging in green exercise even as part of a regular routine like walking, jogging, or riding a bike outside are numerous and not often considered or well-known.

GREEN EXERCISE: THE BENEFITS

Here are some of the benefits of exercising outdoors: (1) Studies have shown that exercising outdoors is associated with even greater stress and blood pressure reduction, compared to similar exercise performed indoors. (2) Outdoor exercise may make you push yourself further. Yes, of course, you can push yourself in the gym. You can listen to your Peloton instructor and crank the resistance up to fifty. You can also certainly increase the incline on the treadmill too. Or you can talk on the phone the entire time and binge watch a show, barely breathless. But when you are outside and faced with a hill on your bike or under your feet, you have no choice but to go up. (3) Sunlight exposure is the best and most natural source of vitamin D. Of course, there are risks related to excessive sun exposure, like skin damage and cancer, but in moderation, exposure to sunlight is the best source of vitamin D and you will get plenty by exercising outdoors. (4) Mental health benefits from outside exercise abound: being in nature; seeing wildlife or birds, trees, and plants; and breathing fresh air all contribute to renewed focus, calmness, and further stress reduction. Something that we all need! With most of us working in offices in front of screens all day, running on a treadmill with yet another screen in front of our faces may be a physical challenge but compromises the mental respite aspect of exercise.

There are also data to suggest even further potential benefits of exercising outdoors near the sea. I know that many are land-locked and don't have regular access to a body of water, much less the ocean. But there is nothing like the vastness of the sea to bring a sense of calm and clarity. The health benefits of salt air have long been touted, not to mention the fact that ocean breezes often mean cleaner, less polluted air to breathe. Even though there is very much a winter where I live, I surf on weekends in the summer, or as I call it my elongated summer: May through October. I'm not very good, since I learned as an adult. But the worst day of surfing is still a great day. Even when I am looking out on the water about to get in and see that the conditions are looking kind of crappy, I will get in just to be out there and get wet. Paddling out against the incoming waves, sitting on the board, feeling how the wind and the tide envelop me are transformational experiences before I have even caught my first wave. Not to mention the fact that the ocean is basically the *only* place I can go where I get to leave my phone behind!

Nature is bigger than all of us. It can put things in perspective and reset expectations. The mental and spiritual health benefits of being surrounded by it cannot be overestimated.

ALCOHOL

Alcohol intake is the only specific dietary factor that we know increases the risk of breast cancer. Heavy alcohol intake, defined as more than seven drinks per week by the CDC, is associated with a greater risk than moderate intake (seven or fewer per week), which, in turn, is more risk than none. We've estimated that a woman who has two drinks a day may have a 15 percent risk of getting breast cancer, compared to a risk of about 10 percent for someone who does not drink. While that statistic might seem significant, we should also keep in mind that two drinks a day is a lot—fourteen a week or more. Alcohol in low moderation (say, three to five drinks a week) may only contribute a minuscule amount of increased risk of breast cancer, approximately 1 percent. Nevertheless, minimizing alcohol intake is one step a person can take to decrease breast cancer risk. If someone wants to do *everything she can* to control or reduce risk of recurrence after a breast cancer diagnosis, not drinking at all is one concrete action item to take. Unless addiction makes it hard for you to drink in moderation, I don't advocate complete deprivation, as a doctor or a human being. You shouldn't need me to tell you that it's okay to have a drink! But if you do, as with Sam in the beginning of the book, I feel that an occasional drink can be a healthy part of celebrations and enjoyment and does not need to be avoided altogether. I drink in low moderation. I won't hesitate to have a cocktail on a Saturday night (especially if I'm offered a margarita!), but I can go days or weeks without drinking any alcohol. At this level, I don't really think about a tangible increased risk to myself or my patients. It's also true alcohol *may* have some benefits related to other aspects of health. Let's not forget that the number one killer of women in America is *not* breast cancer, it's heart disease. Red wine has resveratrol, which may benefit heart health, although the jury is still out on that one. If you're looking for guidance, it's wise to review what you are personally most at risk for, based on your own health and your family history, with your doctor and make an educated decision about alcohol intake based on that.

Diet, exercise, weight control, and alcohol intake are all difficult to optimize. They require discipline and consistency, which can be hard to maintain in our crazy, busy lives.

And if weight and alcohol intake are the two main lifestyle factors that can contribute to breast cancer risk, how do we maintain a healthy body weight when, as we age, biological forces start working against us? Menopause is associated with weight gain. Sometimes that weight gain is on top of prior weight gain related to pregnancies, sedentary lifestyle, or just being too busy or too tired to exercise.

I have one patient who gained a healthy thirty pounds with each pregnancy but only lost twenty after each child was born. She has three children and told me to do the math: she put on thirty pounds over the last ten years of childbearing, exacerbated by working at a full-time job and not having time to prepare healthy food or exercise at the end of her exhausting day. Newly diagnosed with breast cancer, she is petrified that further weight gain related to cancer treatment or oncoming menopause, or both, will make it impossible for her to achieve and sustain weight loss. Weight loss that is now more important to her than ever given that it would now potentially lower her risk of breast cancer recurrence, on top of all of the other overall health benefits. This is an extremely common scenario among the women who come through my office every day. So how does a person optimize her quantity and quality of life related to lifestyle factors either with or without a breast cancer diagnosis?

The first step is to self-prioritize, at least to some degree. For many women it's in their DNA to take care of everyone else *first* and then themselves. But that is not sustainable if you want to optimize your own health. It's recommended that we get an hour of exercise, four or five times a week, as a start. Every day if possible. How do we rally to do this when we are exhausted? More on the importance of sleep later! But part of it is just committing to make a change. That can start with small changes. The patient I wrote about above did something interesting to kick off her weight-loss journey. She lives in a suburb of New York and would have normally had some of her follow-up treatments—like radiation, which was every day for four weeks—closer to home rather than commuting into the city. But she made a different choice. She decided she would take the train into the city every day and then walk the fifty blocks from the train station to our hospital, incorporating an almost three-mile walk into her treatment regimen. By the end of her radiation, she had lost five pounds already and was well on her way.

Weight loss should be programmatic to be sustainable. For those who have cycled through every single weight-loss program out there, the new GLP-1 agonist medications (Ozempic, Wegovy, and others) are incredible new tools to facilitate weight loss and should be considered. I know that some people are reluctant to start these medications out of fear of side effects. Many patients also ask, "When would I stop?" I respond by saying *all* medications have side effects, and in the balance, the pros have to outweigh the cons. While one can discontinue a GLP medication, often the weight will come back. We are learning more and more about these revolutionary medications and both their potential long-term risks and benefits. There is some thought that between the loss of fat stores and the many hormonal and inflammatory agents they generate, the GLPs might decrease everything from cancer risk to heart disease, and even Alzheimer's disease and other forms of dementia. I add that for many battling weight gain or obesity, the number of medications they will end up being on related to obesity have side effects of their own, and they'd never be able to stop those medications remaining overweight.

Recently, one of my patients came in fifteen months after breast cancer surgery having lost over sixty pounds on Ozempic, almost one-third of her original weight of two hundred pounds. She looked and felt incredible. She asked me to look at my original notes from the first time I saw her.

"Look at the medications I was on. Read them to me," she requested.

There were five: Crestor, for high cholesterol. Glipizide, for borderline diabetes. Losartan, for high blood pressure. Labetalol, also for blood pressure and her heart, as well as Zoloft, an antidepressant.

"Ask me how many of these I am on now. Ask me."

"How many?"

"Zero!" she yelled out gleefully, holding up her hand for a high five.

That one intervention probably added years and incalculable

quality to her life. She is off five medications and now on only one, Ozempic. In my opinion, and in hers, it was a good trade.

For patients like her and so many others, a breast cancer diagnosis can be a reckoning. An opportunity to improve lifestyle factors that, well, could use some improvement. My friend and oncology colleague says that in the wake of a breast cancer diagnosis he advises his patients to be mindful of the "three *ws*": weight, wine, and walking. I would add a fourth *w*: working out. Walking is important to avoid a sedentary lifestyle, so get those steps in. But more vigorous exercise is important too. If your doctor gives you the go-ahead, don't be afraid to go hard.

The bad news is that, even taking in all the factors we've discussed above, there is no magic bullet or potion to completely eliminate risk. Reducing alcohol intake, reducing weight, exercising, and eating healthfully are all things one can do to reduce the risk of getting breast cancer, reduce the risk of recurrence, and improve one's overall health. But for a woman with breast cancer, these lifestyle interventions should not come from a place of blame or self-flagellation. You do not deserve punishment for something that was not your fault to begin with. And if you reframe your relationship to healthy living as a declaration of love and support for yourself, you just might find these new habits are pleasure not punishment.

HOW ARE HORMONES INVOLVED

ARE THEY A GODSEND OR SHOULD YOU PROCEED WITH CAUTION?

JESSICA G., account manager: When I was diagnosed with breast cancer, I had been on hormone replacement therapy for terrible menopausal symptoms for about eight years. I started to experience menopause much earlier than the rest of my friends, at about age forty-five. I had been getting hot flashes and night sweats for months. They would wake me up in the middle of the night. In the morning, I would be exhausted and could barely get my kids off to school and drag myself to work. It started to compromise my performance in the office. I am an account manager for a financial company, and I look at spreadsheets and portfolios, and often sit for hours at a time in an office at a computer. When my supervisor caught me, for the third time, with my head down on my desk taking a nap in the middle of the day (because I had slept like four hours for the last three nights in a row), I thought I would be fired. I just couldn't function and had to make a change. I saw my doctor and discussed my options. She started me on hormone replacement, and within a month I was a different person, sleeping through the night again, and gradually recovering from the prior year of extreme sleep deprivation.

I knew that there was a small increased risk of breast cancer associated with taking hormone medication, we had discussed that, but I was willing to take that risk and promised myself I would be careful and diligent about monitoring. I know some of my friends won't take it because they have other risk factors for breast cancer, like a family history. They weren't interested in adding to their risk even a little bit. Also, some of my friends didn't seem to be suffering like I was! Their symptoms were annoying, not incapacitating, as I felt mine were. Like most women who get breast cancer, I didn't have a family history. I felt comfortable taking a small risk for the incredible improvement in quality of life that I desperately needed. In the end, as my doctor reassured me, there is no way of knowing if the hormones actually *caused* my breast cancer. I might have gotten it anyway. Now that I have been diagnosed with breast cancer, I must go off the hormones, since they are feeding my cancer. I understand that. But it's interesting: it's eight years later, and my symptoms are not as severe. I am doing okay. For me, taking the hormone therapy at that time made all the difference during the period of my life I needed it most, and I don't know how I would have gotten through that difficult perimenopausal period without it.

My twenty-five-year-old daughter has been on the birth control pill for about five years. We talk about the small added risk of breast cancer to her, especially now that she has a family history: me. She also believes that the very small added risk of breast cancer related to the pill is outweighed by the benefits to her she has discussed with her doctor. And she knows that she can go off at any time. While that won't reverse the small risk she has accrued, it won't increase further. We are good with our decisions and know that they are totally ours to have made and to make in the future.

DR. PORT: The previous chapter discussed a bit about how hormones from fat stores increase the risk of getting breast cancer for women who are overweight or obese. But what about hormones from outside sources: medications and pills we can take?

Hormone replacement therapy (HRT) was developed to quell menopausal symptoms and effectively does just that. Menopausal symptoms can vary in kind and severity from person to person but include any or all of the following: hot flashes, which can manifest as night sweats; sleep disturbances; vaginal dryness; decrease in libido; weight gain; hair thinning; and perception of brain fog, to name a few. Ask a few women how they were affected by menopause and you may get one of the answers above, some combination, or all of them. That doesn't even include the slew of other, more vague symptoms that women also attribute to menopause, like headaches or a decrease in energy level. Whether you are taking care of a family, working a job, both or neither, aging is tough. But as my father would say, in his best impression of a Borscht Belt comedian: it's better than the alternative. Menopause is part of the normal aging process in women starting around age fifty, although it can range from earlier to later, and symptoms can be anywhere from mild and short-lived to severe and long-term.

Since forever, so many women have suffered in silence from symptoms related to menopause that can be quite disruptive. As Jessica described, sleep disturbances—which can be related to fluctuating hormone levels or night sweats that wake you up—severely affected her quality of life. But it's also important to know that long-term poor sleep can be detrimental. Without getting into too much of the science, "sleep hygiene" is a key but often overlooked component of overall health. When you are awake, even when you think you are "relaxing," say sitting quietly on a lounge chair by the beach somewhere, your brain is receiving, processing, and responding to so much information and so many stimuli. These activities all consume energy. It's estimated that while awake, your brain consumes about 20 percent of the body's energy, while making up only 2 percent of the body's weight. When you sleep, both your body and your mind power down. Your body can then devote all efforts to performing reparative and restorative functions. As you heard in the introduction, my patients feel exhausted after surgery because their bodies are working double time to perform normal activities *and* repair significant wounds from surgery. But even when you aren't recovering from surgery, there are critical reparative and restorative processes that happen during sleep such as tissue repair, immune system surveillance, and cognitive functions like memory consolidation. All of your body's efforts and energy can be channeled toward these processes while you are asleep and not thinking, reacting to your environment, or running around. So, managing menopausal symptoms that include sleep disruption is important for both quality of life and overall health.

Women and their doctors often turn to HRT to cushion the blow of the changes associated with menopause when they are severe enough to affect quality of life. Hormone replacement usually involves a combination of estrogen and progesterone, taken together in the form of a pill, a patch, or gel that is applied to the skin. Like the hormones we produce in our own bodies, known as *endogenous* hormones, the hormones we take from outside sources, or *exogenous* hormones, have varying effects on different parts of the body. In addition to alleviating menopausal symptoms for some women, HRT can have other positive effects. But like all medications, there are risks and side effects too. And that's because when we take medications they go everywhere in our bodies and can have positive effects in some tissues and negative effects in others. For example, HRT causes an increase in bone density and stimulates bone growth, which is a good thing. But in breast tissue, that same stimulating effect can lead to cell changes and ultimately the development of breast cancer.

To be clear, the increased additive risk of breast cancer with HRT is quite low: the average woman's risk of getting breast cancer over her lifetime is 10 to 12 percent. HRT might increase that to 12 to 14 percent. It has other risks, too, like blood clots. Interestingly, estrogen alone doesn't increase the risk of breast cancer much at all; it's the combination of estrogen and progesterone. Of course, the next obvious question is: Why not just give estrogen alone? Estrogen alone, also called unopposed estrogen, increases the risk of uterine cancer, so it's not a good idea. So, estrogen alone can increase uterine cancer but not breast cancer, while estrogen and progesterone together increase the risk of breast cancer but not uterine. I know, it sounds like these are two equally bad choices. (And, by the way, why is there not a door number three? A choice that doesn't increase either of these cancer risks?!) But for women who need to handle severe symptoms, the benefits outweigh the risks. Above all, it's important to assess your own risk factors and go from there.

It's also important to know that there are different kinds of HRT and different modes of taking it. Pill, patch, pellet, and cream forms can all deliver hormones that get absorbed and go everywhere in the body. There are different doses, too, and no matter how low the dose, the risks are small but real. If you don't have breast cancer, taking HRT should be an option if you are suffering from any (or all!) menopausal symptoms. To reiterate, the added risk of breast cancer is small. But once a woman has developed breast cancer, which are most commonly hormone-fed cancers, giving supplemental hormones that could feed cancer cells is no longer considered a safe option, with rare exceptions. Why? First of all, many women diagnosed with breast cancer are recommended to go on *anti*hormonal therapy to prevent breast cancer recurrence, and we can't both give and block hormones at the same time. Together, the two would undermine each other, and ironically, since some of the side effects, like blood clots, are the same for both pro- *and* antihormonal agents, taking them together could magnify those dangerous risks.

Many women who come to see me tell me they were taking a "safer" form of hormones called bioidentical hormones and ask whether they still have to be stopped. Bioidentical hormones are purported to behave in the body like hormones (biologically identically), but are from a more "natural" source and therefore lower risk. They are marketed as safer or seem safer as many are derived from plant sources, also called phytoestrogens.

Unfortunately, many of these products are more like supplements than actual medication and are therefore untested and unregulated. The thing to remember if you are taking these is that when we put something—anything—in our bodies, we don't get to pick or choose where it goes. We don't get to direct a medication to help with our hot flashes but stay away from our breasts. Medications and treatments get into the bloodstream and circulate everywhere. So, if bioidentical hormones *are* helping with menopausal symptoms and feel effective, they will increase breast cancer risk, too, to some degree. There is no free ride here, despite what anyone might claim. Because many of these bioidenticals are unregulated and don't have to prove effective dosage or consistency between doses, many will be mere placebo pills filled with dried-up plants, or even rice powder among other fillers. Personally, I would rather take a medication that *is* approved and manufactured in a regulated facility, so I know what I'm getting: the risks and the rewards.

October 2025 brought an exciting and interesting development related to HRT. The FDA removed the "black box" labelling of common hormone replacement medications that had been placed as a warning against side effects. It's important to know that a black box label on a drug implies the highest level of risk related to adverse side effects. For HRT, this warning was mainly about breast cancer and cardiovascular risks, based on results from a study published in 2002 called the Women's Health Initiative (WHI). The study showed an increased risk of both invasive breast cancer (25 percent) and cardiovascular events (almost 30 percent), but a *lower* risk of fractures and other cancers, such as colorectal and ovarian. This labelling definitely deterred many patients from taking HRT (and their doctors from willingly prescribing it) due to the outsized perception of risk. And in fact, in the wake of this study, the incidence of breast cancer in America *did* go down, most likely related to cessation of use by many women. What many did not realize is that this defining study on which these risks were determined, and on which the decision to place this extreme warning level was based, had significant limitations. For example, the majority of women in the WHI study were sixty and older, and heart disease and breast cancer are much more common in women sixty and older, and in fact increase with each passing decade. Most women who initiate HRT start it much younger, like Jessica did, when they are actually first experiencing menopausal symptoms. The risk of developing breast cancer and cardiac disease that the study reported may be related to the later starting of hormone replacement, at a time when women were at increased risk for these side effects anyway. It's possible that for younger women, predominantly in their late forties and early fifties, HRT would not be associated with as dramatically increased risks as what was seen in the WHI study.

I am a big proponent of hormone replacement for quality of life and overall health, and the benefits most likely outweigh the risks for those at average risk for breast cancer who suffer from menopausal symptoms. For those at increased risk for breast cancer, hormone replacement should also not be completely off-limits but should be considered on a case-by-case basis. This leaves the only remaining group, who should not take it, being those diagnosed with breast cancer themselves. There are also newer, extremely exciting formulations that may have no added risk of breast cancer that are being tested and studied. Hopefully, more on these soon.

What about birth control pills and other hormone-based contraceptives like Jessica's daughter was taking? They, too, add hormones to our systems. Because birth control pills both add hormones and suppress our own hormonal surges, for a long time it was controversial whether or not they did increase the risk of breast cancer, with conflicting evidence pointing in different directions. Until 2017, when a large and fairly definitive Danish trial, which tracked 1.8 million medical records, showed that use of hormone-based birth control in women between the ages of fifteen to forty-nine did increase risk of future breast cancer. The risk increased the longer the utilization, with very little added risk among those who used it for less than a year, compared to a significantly higher increase among those who were on it for ten years or longer. The study concluded that the increased risk among those who took birth control meds for ten years or longer was almost 40 percent. Now wait: before your jaw falls on the floor and you run to flush your medication down the toilet, let's explore exactly what that means. If the average woman's risk of getting breast cancer is 10 percent over her lifetime, increasing that risk by 40 percent does not mean her risk is now 50 percent. A 40 percent increased risk means adding 40 percent of one's baseline risk, 10 percent. If you do the math, 40 percent of ten is four. So the *actual* added risk is 4 percent. The Danish study showed that women who were on birth control for more than ten years had an approximately 14 percent risk of getting breast cancer over their lifetimes compared to women who were on no hormonal birth control at all who had a 10 percent risk. Importantly, birth control pills have other effects that are beneficial, aside from preventing unwanted pregnancy. While birth control medication increases risk of breast cancer, it actually *decreases* risk of ovarian and colorectal cancer. That perspective here is crucial, since understanding these numbers is critical for making the kinds of decisions that Jessica and her daughter were able to make in regard to the hormones they were taking. So, are hormones a godsend, or should we proceed with caution? The answer is probably a little bit of both, depending on the individual and her other health factors.

Hormonal factors that can either increase or decrease the

risk of breast cancer are also related to childbearing and reproductive history, in addition to the kind of hormones you can take as a pill or a patch. The hormonal cycle in and of itself, which exposes breast tissue to rising and falling hormone levels, plays some role, albeit small, in breast cancer risk. That's why, as a general rule, the more menstrual cycles a woman experiences over her lifetime, the greater the risk of breast cancer. The average age of a girl starting her period, or menarche, is now 11.9 years old. For that reason, a young girl who experiences menarche at 9 or 10 years old will be at higher-than-average risk for breast cancer based on this factor alone. In 1970, the average age of menarche was 12 to 13 years, while now, fifty years later, that age has dropped to 11.9. I know this may not seem like a big drop, but it is a significant trend indicating many more girls and young women are getting their periods a year, two years and even three years younger than the average. Interestingly, before the 1900s the average age for a girl to get her period was almost seventeen. Why is this drop in average onset of age interesting or even concerning? This long-term trend in the decreasing average age of menarche is one of the explanations that has been put forth for the increased incidence of breast cancer in young women. Similarly, the later age of menopause onset, the higher the risk, for the same reason as early menarche: later menopause means more years of menstrual cycles, and again, hormonal cycle exposure to the breast tissue that goes with it.

In terms of other reproductive factors, we also know that not having a child is associated with a slightly higher risk of breast cancer than having children at age thirty or older, which, in turn, is associated with a higher risk than having children at a younger age. This is mostly likely related to the disruption of hormone cycles that happens during pregnancy. It's true that the body is a hormonally rich environment during pregnancy, but the ups and downs of menstrual cycles, which cause the breast tissue to change cyclically, does not occur in pregnant women. By sparing the body a few extra menstrual cycles, pregnancy at a younger age is protective against breast cancer. Finally, breastfeeding has also been shown to be marginally protective. Again, the specific mechanism is unclear. We're not sure if it's related to the continued disruption of cycling that breastfeeding often leads to or if the milk ducts, where breast cancer usually starts, are "cleaned out" in some way with the flow of milk. What we do know is that when comparing large numbers of women who do and do not breastfeed, after controlling for other factors, the breastfeeding women had an overall marginally lower risk than their non-breastfeeding counterparts.

To state the obvious, most of these hormone-based risk factors are completely out of our control. We can't control when our bodies decide to have their first period or their last. To a certain degree, we can control and make decisions regarding childbearing: to do or not to do and when. But there are so many other factors that play into making these decisions: relationships, life circumstances, and most importantly, personal choice. I certainly wouldn't recommend that a woman factor in the increase or decrease in risk related to breast cancer when making such serious and dramatic life choices. Please don't decide to have a baby at twenty-five just to reduce your breast cancer risk because in the end, thankfully, the increase or decrease in breast cancer risk related to most of these hormone-based factors is quite small in magnitude. If the average woman's risk of getting breast cancer is 10 to 12 percent over her lifetime, the worst combination of these hormone-based factors probably only increases her risk by 1 or 3 percent.

SUPPLEMENTS

IS THERE A SECRET POTION TO GIVE YOU THE EDGE?

TERRI T., high school teacher: I was diagnosed with breast cancer after I felt a large lump in my left breast. I'm not going to get into it, but I had a lack of faith in the value of getting mammograms. I did yoga. I was vegan. I took immune-support supplements. I thought that my lifestyle would protect me from any form of life-threatening illness. When I was diagnosed, my first thought was that I would take control and cure this cancer my way. I started taking about forty different supplements that were purported to have anticancer properties. Despite my doctors' grim warnings, I did that for three months. Then, when my cancer doubled in size, I came in to see Dr. Port. At that point, I was pretty desperate. I was ready to talk about mainstream and standard of care approaches. It also dawned on me that perhaps, given the rapid growth, all the supplements and "boosters" feeding my immune cells might have been feeding my cancer too. She made me go off all the supplements, and I agreed. I had tried it my way, and it didn't work. In addition, apparently, some supplements thin the blood and increase the risk of bleeding at surgery, while others affect liver function, and a lot of the anesthesia medications I would need for surgery are broken down by the liver.

It's been a year since I started my treatment, and I'm doing well now. I have gone back on some of the stuff I was on before, specifically some of the vitamins, but I now know that they cannot cure my cancer. I am also very up-front with my doctors about what I am taking so they don't interfere with the meds that my doctors have me on that *do* cure my cancer or at least prevent it from coming back. My advice is to be cautious about an "alternative" approach, meaning alternative to mainstream treatments. It's soooo tempting to think you can take control and cure cancer your own way, but in my case, it was a huge setback.

DR. PORT: Many of my patients channel their terror of recurrence into extreme lifestyle changes with the illusion that these will impact their health in a positive way. They come into my office six months after treatment is done declaring they have become vegetarian or gone vegan or given up red meat or sugar or are running six miles a day or meditating. For many, the allure of thinking that changing just one thing will improve outcome is too tempting to resist. And there is plenty of misinformation out there that will support the desire to do so. Many of my patients are on all kinds of supplements and vitamins even before their diagnosis: multivitamins, echinacea, ginkgo biloba, turmeric, and the list goes on and on. They take them to address a variety of different health concerns from arthritis to cognitive function—and, of course, cancer prevention. When it comes to supplements or additives that are purported to be cancer busters or immune boosters, there are countless vendors who are happy to sell something with no proof or validating studies that it provides any benefit. The false promises of getting an edge over cancer can be very seductive.

Because so many of my patients come into my office with a long list of different additives they are taking, I'm pretty well versed in the wide variety of supplements out there, but there were some on Terri's list I had never heard of, including one called turkey tail, which I was fascinated to learn is not, in fact, the tail of a turkey, but derived from a mushroom. The first thing Terri said to me was that it was very important for her to feel that she was in control of her care and her cancer. I thought, *This is going to be a tough one.* But the reality is, patients do need to know that they are in control, to the degree that this is possible over any outcome. They *do* make their own choices. We, their doctors, are just purveyors of information; technicians who perform surgery; and oncologists who deliver medications and treatments that prevent or reduce the risk of cancer coming back. If you really want to, you can stay on all of the supplements you want, regardless of our advice against it.

But here are the reasons why it's important to heed Terri's advice to be cautious about an alternative approach. The world of supplements is vast. The approximately $180 billion global industry of vitamins, supplements, IV treatments and injections is almost entirely unregulated, which means that there is very little oversight related to validation, effectiveness, quality, or quantity. Let me break this down. First, the purported benefit of most supplements—that is, whether they actually do what the label says they do—is speculative at best, in most cases. Rare exceptions aside, the makers and manufacturers of supplements have never performed the kind of scientific studies to prove these benefits (or equally as important, delineate side effects). Anyone can make a supplement and start selling it. But for any medication to reach the point where I'd be offering it to a patient, it would have to go through extensive studies, including safety studies, dosage studies, and often a randomized controlled trial, with treatment and placebo arms, to prove that the treatment was more effective than a placebo. This is the gold standard for determining effects, positive and negative, on a patient. Without clinical trials, how can we know if someone's joint pain improving while taking ginkgo biloba is related specifically to that supplement or a potential placebo effect? The reason why producers of these supplements don't test them is because they don't have to. It could take millions of dollars and many years to conduct the kind of rigorous clinical trial to prove efficacy. If we're buying them anyway without proof that they work, why should the manufacturers go through the trouble? Besides, for some (most?) of these supplements, a well-run clinical trial will only demonstrate that they are no more effective than placebo. That would be bad for business.

I have an acquaintance whose family business is a chain of health food stores. A few years ago, she called me about a new probiotic concoction they were looking to sell. She asked if I would be willing to try it, and even perhaps endorse it. Then, she went on to regale me of all the benefits: the purity of the ingredients, the quality of the product, the cleanliness of the manufacturing process.

"I went to the lab myself, Elisa. It's amazing and immaculate!"

"You mean on that specific day?" I responded.

Needless to say, she was not happy with my snarky answer. But I was serious. Why would she believe a lab, whether in the US or abroad, would be held to an adequate standard of safety, cleanliness, consistency?

I asked a lot of questions she was not prepared to answer. By the end of the conversation, she asked me, "Why are you so skeptical?" To which I answered, "Why are you *not?*" A lot of supplements are made in independent labs around the world then sent to the United States for slick packaging and marketing, which, it turns out, is exactly how that product got to her store. Most people, when put on a medication, want to know what the potential side effects might be. But ask any person on supplements what the side effects are, and you will be met by either a blank stare. Or they will answer, "Nothing. There are no side effects." That's just not true. Anything and everything that you ingest, inject, or smear on—anything that has a biological effect on any level—has the potential for side effects. Nothing is all good and no bad. The only things that have no potential for side effects are placebos because they are absolutely biologically inactive. Since no structured studies have been done for the vast majority of supplements, we are unable to accurately determine what side effects they come with. My advice is that we should, at the very least, hold supplements to the *same* standards as over-the-counter and prescription drugs. You should ask the purveyor what the side effects are; if the answer is that they don't know or that they haven't been studied, then walk away.

Speaking of placebo, let's talk about quality. In 2015, the attorney general in New York State at that time, Eric Schneiderman, in response to reports and concern regarding the safety of supplements, initiated an independent investigation of a broad array of supplements across a broad array of purveyors. Investigators simply walked into these stores, randomly selected items off the shelf, and took them to an independent laboratory for testing. The results were equal parts astounding and disturbing. Four out of five items tested had absolutely no content of what was on the label. The echinacea had no echinacea. The chondroitin sulfate had no chondroitin sulfate. The turmeric supplement had no more than what you would use to season food. But the bottles weren't empty. So, what *was* in there? In short, filler: flour, rice powder, even dried houseplants. This investigation and its discoveries resulted in a cease-and-desist order to these major purveyors. Unfortunately, it did not result in the changing of any laws, regulations, or standards governing the quality of supplements, production, distribution, or sales. Beyond the danger of ingesting these by-products, I find it disheartening that so many people are getting nothing for their money. Most of us can't afford to waste up to hundreds of dollars a month on something we hope will help us but is, very likely, a placebo or worse.

Finally, there is the quantity factor. As discussed earlier, when clinical trials on medications are done, there are different phases. The earliest phase is to establish basic safety, and appropriate dosing. Trials are done with patients on escalating doses of a medication in order to reach best levels of effectiveness, weighing optimal positive effects against side effects. Determining dosing is a critical part of establishing safety *and* efficacy. Without clinical trials with escalating dosage, there is no way to determine comparable effects or lack thereof. Almost no supplements go through this "phase" (or any phase) of clinical trials. The dosing we see on their packaging is basically guesswork and completely unvalidated. More concerningly, what the attorney general investigation showed, as have many since, is that there isn't even consistency across batches in the dosage one might purchase. The bottle of vitamin E capsules you bought in May might or might not have 15 milligrams in each capsule, as it says on the label. When you go to replenish your supply in July, the same bottle with the same label may have something totally different in it. One batch may have 15 milligrams in each capsule; the next batch may be filler.

All of this to say, the major gap in quality control in the multibillion-dollar supplement industry has led to well-known safety issues. The most glaring manifestation of this can be seen in the number of yearly emergency room visits related to supplement use. Supplements tend to be broken down and metabolized by the liver, and approximately 25 percent of emergency room visits related to liver toxicity can be attributed to supplement use. Additionally, lots of the supplements aimed at bodybuilding and increasing muscle mass have caffeine and other stimulants in them, leading to people who take them feeling a racing heartbeat or shortness of breath, even fearing they might be having a heart attack. Finally, even the most seemingly innocuous supplements can interfere with or compete with other prescription or over-the-counter medications, leading to magnified side effects from either or both, or cancellation of a medication's desired effect.

There are, of course, exceptions to the rule. When it comes to taking supplements or vitamins where there is a recognized bodily need or potential deficiency, supplements can be helpful, as long as they're legit. The great historical example of vitamin deficiency comes from stories about sailors who developed scurvy on long journeys at sea, where food supplies were low and there was often no access to fresh fruit or vegetables. When the sailors reached shore and ate oranges, tomatoes and other dietary elements rich in vitamin C, rectifying the deficiency, the disease was magically cured. Today, people who have malabsorption conditions may need to take vitamin B12. Someone who doesn't get enough sunlight should take vitamin D. Adequate calcium is necessary for bone density, especially for women as we age. Zinc is necessary for hair growth, as I learned when I was a surgery resident, working a hundred hours a week, subsisting on mostly M&M's from the hospital vending machines on my grueling overnight calls. I would buy the jumbo pack of M&M's and ration them over the course of the night, finishing the last few at around 4 or 5 a.m. I joked with my fellow residents that M&M's do provide all the nutrients one needs: the red ones were the meat group; yellow were butter and dairy; the green, of course, were the vegetables. The M&M diet went on for months, every other night on call. The jig was up with my ridiculous vending machine diet when I got home one night after thirty-six hours in the hospital, took a much-needed shower, got out, dried off, and found the entire center section of my left eyebrow in my towel. Evidently, zinc, which is necessary for hair growth and retention, is *not* an ingredient in M&M's. Clearly, I needed to diversify my diet and get back to at least some healthy eating.

Whether or not supplements are able to compensate for vitamin and elemental deficiencies, rather than food, has not been well studied. Again, if there is no actual vitamin D in the vitamin D capsules, it won't help. In summary, very few people need dietary supplements or vitamins above and beyond what one can obtain from a healthy diet. However, the need for dietary supplementation is certainly worthy of a discussion with your doctor.

What about patients with cancer like Terri? Well, the same goes, but there is an added concern. Some of my patients, even those who have no intention of treating their cancer with supplements like Terri did, start taking supplements that have been suggested to them by their friends who have been through cancer themselves, by their doctors, or by articles they've found online. The desperation to get "an edge" can manifest itself by taking immune boosters; cell growth promoters to enhance wound healing from imminent surgery; or supplements marketed as "cancer-fighting" formulas. Many of my patients seek out these products and start taking them almost immediately after diagnosis to "get ready" for surgery or build up their immune systems. The problem, which I point out when they come in to see me for the first time, is that when they take these products—by pill, by capsule, or IV drip— *they* don't get to pick and choose where this stuff goes. They have not had surgery, so the cancer is still in their body. Even after I perform surgery, there is always the possibility of microscopic cancer cells left behind. Is it possible for the immune booster to reach the tumor or residual cancer cells, and give *them* a boost? Theoretically, could "wound healing" supplements promoting cell growth work the same way on cancer cells? It's usually at this point during the consultation when the patient's jaw drops. "I had never thought of it that way" is the typical response. "Don't worry," I respond. "Probably no harm done with such short-term usage. Just get off them." Then I think to myself, *They were probably placebo anyway.* When deciding what to take and what not to take, especially in or around a cancer diagnosis, you'll want to remember that there is always a potential downside. Proceed thoughtfully and after full discussion with and disclosure to your doctors.

STRESS

MY DIVORCE GAVE ME BREAST CANCER, AND OTHER MYTHS

ANDREA K., engineer: It had been a really bad couple of years leading up to my breast cancer diagnosis. My mother died, and my then-husband Jake and I had been having issues for a long time. He lost his job and, for whatever reason, couldn't seem to find another one. My company, a tech start-up, was exploding and had me working insane hours. I was barely managing responsibilities at home with our two kids *before* the pandemic. Every day was a struggle to keep all the balls in the air. I cut my husband some slack because I knew it was depressing for him to be out of a job and having such a hard time finding another. But it didn't seem to me he was even trying that hard. I was on the verge of asking him to move out and get his shit together when COVID hit. Then, since everything was so unstable, I thought we should be together under one roof as a family, for the sake of our kids. But the pandemic and the total halt in his industry, real estate, sort of gave him permission to stop looking for work. The problem was I was still working like a maniac, just from home. My company went into crisis mode with COVID, but we had total capacity remotely, and we were expected to be available 24–7. I was burning the candle at both ends, and my husband wasn't picking up the slack. I became super resentful. It wasn't his fault that he couldn't find a job, but the least he could do was help out at home—which he did not. I get it, he was depressed. And depression makes you feel like not doing anything . . . certainly not laundry or helping the kids with their math homework after a day of online school. But I thought we were a team. I lost my teammate, and I felt abandoned. So did he, I am sure. We were two people living together but feeling completely emotionally and logistically abandoned by the other. While we tried to remain civil to each other in front of the kids, it was understandably tense between us around the house.

The combination of my job's intensity and insane amount of stress along with the problems in my marriage that became amplified by the pandemic were affecting me for sure. I was run down. I felt totally unhealthy. We weren't eating well, ordering in pizza or Chinese food every night. My sleep was constantly interrupted by calls from overseas clients. And I certainly had no time to exercise. I just don't think it's an accident that I developed breast cancer after this terrible, stressful two-year period. I know it's hard to say for sure that my life circumstances gave me cancer, but given that this is my reality, I am committed to doing what I can to manage my stress level. I am making adjustments at work. Now that my husband

and I are truly separated, he has full responsibility for the children when they are with him. I miss them terribly when they aren't with me, and it's lonely at home, but at least he is stepping up with the kids, if only because he has to. Maybe we would have worked it out, if he'd been able to take responsibility sooner. It's so sad. Even if trying to manage my stress levels doesn't impact my quantity of life, it will definitely affect my quality of life. So that is what I am trying to do. I would urge anyone going through a breast cancer diagnosis, especially against the background of other life stressors, to do the same if at all possible.

DR. PORT: I recently read something posted that said, "90 percent of the things I worry about never happen. Worrying works." It's true that for some things, worrying can be preventive: if you worry about being hit by a car while crossing the street, you might pay more attention and be more careful to look both ways before you cross. As it relates to breast cancer, if worrying about getting it leads one to channel that worry into a lifestyle that is associated with lower risk, then it might be helpful and productive. For a woman with breast cancer, worrying that it might come back might lead her to be very conscientious about completing treatments and taking recommended medicines to reduce the risk of recurrence. But outside of these channels, worrying does not "work" to prevent breast cancer or its recurrence for someone already diagnosed. I felt comfortable reassuring Andrea that the extremely stressful period she had endured did not cause her breast cancer.

The cause-and-effect relationship between stress and cancer

has been a controversial one for a long time. There are many studies too complicated to dive into here showing that many of the "stress" hormones—cortisol, insulin-like growth factor (IGF), adrenaline—can stimulate cancer cell growth or promote cancer spread or metastasis. The theory is that sustaining high levels of these circulating hormones as a result of chronic stress could have a causative effect on the development of cancer or enhance the growth of cancer. There are limited animal studies demonstrating, for example, that mice with tumors placed in a stressful environment, such as isolation, experience greater growth and progression of their tumors compared to mice that are not stressed. In the end, for every study linking stress to cancer, there's another showing no such connection. For example, one large British study of more than 100,000 women did not show a consistent connection between a person experiencing a significantly stressful event during the preceding five years and the development of breast cancer. Similarly, a fifteen-year Australian study showed no correlation between acute or chronic stressors and the risk of breast cancer. Looking at the data, it's important to remember that the relationship between stress and cancer can be indirect too. For example, stressful life circumstances can lead to unhealthy behaviors such as smoking, excessive alcohol intake, or overeating leading to weight gain, all of which are clear and known contributors to increased risk of different kinds of cancers.

One of the main problems with defining the relationship between stress and cancer in an individual is that people may *experience* stress differently. Trying to finish a project on an unrealistic work deadline may make one person distraught. Trying to get to work on time while stuck in insane traffic may feel extremely stressful to another. In addition, episodic stress, like sitting in traffic or meeting tight deadlines, can affect us differently from sustained stress related to long-term traumatic situations or circumstances, such as taking care of a sick relative.

There have been studies performed on people who have experienced long-term, extremely stressful, even life-threatening situations: prisoners of war, Holocaust survivors, hostages. Long-term stressful environments, such as captivity, have been shown to weaken the immune system. There is also some evidence that individuals exposed to the most severe forms of stress for long periods have a slightly higher incidence of some cancers, namely colon and lung cancer. These specific cancers, of course, may also be related to the long-term exposures to severely unhealthy environments, as well as poor hygiene, increased risk of infection, and extremely poor nutrition, so these studies do not conclusively rule in or out a causative link between stress and cancer.

Ultimately, the idea that stress causes breast cancer to develop or progress is also problematic because the *last* thing that a newly diagnosed person wants to hear is not to worry. Worrying is bad enough without having to also worry that you are worried. Or that worrying will make newly diagnosed cancer grow or spread. Nothing could be further from the truth. At the end of the day, the best thing we can do in the face of stress is to be optimistic. Optimistic about one's future, in one's outcome if already diagnosed with cancer, and as a general state of mind. Having an optimistic outlook is often hard in this crazy, upside-down world. However, managing stress is an important part of maintaining an optimistic outlook, and vice versa. Being optimistic that you will survive and in fact thrive after a cancer diagnosis itself can reduce stress.

As Andrea pointed out: while the role that stress plays in

getting or surviving cancer, the quantity of life, may be in question, its role in one's quality of life is not. Committing to lifestyle changes, including managing stress levels, is an important part of how we improve our quality of life, with or without a breast cancer diagnosis. Some of the tools that people implement to effectively accomplish stress reduction are activities such as exercise, cultivating strong, positive relationships, letting go of negative ones, therapy, having spiritual connections, and engaging in meaningful work and hobbies. We all have to find our own ways toward a more balanced, stress-reduced life. It's not about trying to be completely stress-free. That's not a realistic goal. But stress *reduction* is a health goal we should all strive for. Not because stress will give you breast cancer. But because it will improve your quality of life, and that can start today.

THE BREAST LIFESTYLE ADVICE

PATIENTS' WISDOM AS YOU APPROACH LIFESTYLE DECISIONS

CINDY M., environmental lawyer: Dr. Port told me, "We need to stop asking ourselves, 'Was it something I ate?'" It wasn't. She told me she has patients who are vegetarian triathletes who have gotten breast cancer too. And many Cheeto-eating Diet Coke drinkers don't! It doesn't seem fair, but it's true. At least as it relates to getting breast cancer or having it come back. And at least we can stop blaming ourselves and move forward.

TERRI T., high school teacher: My advice is to do something, anything to make you feel like you are living a healthier life. It definitely made me feel like some things were in my control.

MY ADVICE

Taking stock of lifestyle factors that influence overall health—and therefore breast health—is always worthwhile. Maintaining a healthy body weight, incorporating exercise into your lifestyle, and moderating alcohol intake, if you choose to drink, are all interventions that can reduce breast cancer risk, creating a healthier lifestyle and improving quality of life as well.

SCREENING

5

MAMMOGRAMS

HOW TO MAKE SENSE OF THE MOVING TARGET THAT IS THE GUIDELINES FOR SCREENING

JENNIFER L., flight attendant: I'm forty-nine and had been begrudgingly getting mammograms every year since I turned forty. I hated them. They were uncomfortable. They squished my sensitive breasts. Every time they said, "Hold your breath" it felt like an eternity, and I almost passed out.

Five years ago, they found something that needed a biopsy. I freaked out. I was scared out of my mind. When I found out the results were normal a few days later, I basically collapsed in relief. The waiting and the anxiety were so awful that I said to myself, *I can't go through this again. Maybe I'll skip a few.* Then COVID hit, and I did skip a few, like everyone else I knew.

It also felt like the advice on when to start getting mammograms kept changing over the years, which made me wonder if there was even a consensus. I felt like I could make up my own schedule, but my gynecologist urged me to get back on track with regular screenings, so I finally went. When I went back to get my first mammogram in a few years, they found something new, this time in the other breast. *Here we go again,* I told myself. Because I had been through this cycle of waiting and spiraling before, I wasn't as anxious as the last time; I felt like I knew what to expect. Plus, my breast radiologist said, "It's probably nothing."

Imagine my shock and horror when I got a call three days after the biopsy telling me that this time it wasn't a false alarm: it was cancer.

I was in disbelief. "I thought you said it was probably nothing?" I yelled through the phone at the radiologist, definitely shooting the messenger. She replied, "Listen, most biopsy results *are* normal, but yours was not. I am so sorry, but it's very small, tiny in fact: early, treatable, and curable. You are a poster child for why mammograms save lives." Evidently, this was not her first rodeo.

Finding out that I had breast cancer was one of the worst days of my life. The days to follow weren't so great either. Getting that diagnosis changes you forever, on some level. I still think of it in some capacity almost every day. But getting a mammogram didn't *cause* my breast cancer; it picked it up early. Aside from the psychological trauma of knowing I had cancer, the actual surgery and treatment to follow were fairly easy to go through. I keep wondering what would have happened if I had *not* gotten my mammogram. I'm pretty sure that at some point I would have felt a lump, and that would mean it was bigger. When I got my mammogram, my cancer was two millimeters, the size of a pinhead. My doctor told me I might not have been able to feel it on a self-exam until it was almost five times that size. In that time, which could have been months or even years, the cancer cells would have the opportunity to spread, compromising my survival. Even if it would still have been curable, the treatment I would have needed for a larger, potentially more advanced cancer would most likely have been different. My advice is to suck it up and get a mammogram. Mine might have saved my life.

STACEY GRIFFITH, founding senior master instructor of SoulCycle: I am one of the healthiest people I know and didn't realize how important screening actually is because I hadn't gone in years. I was probably already well down the path of living with cancer in my body pre-COVID and didn't even know it. So, when COVID hit and there were no appointments available, I didn't get screened. Then when appointments started to roll back into a normal schedule, I kept trying to schedule one but I would be like, *That doesn't work. That doesn't work either. I can't go then. I can't make that appointment.* So I just kept pushing it and pushing it. A whole year went by after COVID. Two years, or maybe longer, into carrying the cancer around, I felt a lump. I might have avoided all of that chemotherapy and radiation had I caught it earlier by screening, where you get those initial signs that you have something. Pushing those appointments kind of screwed me. That's why I'd love to help get the message out that early detection is key to less invasive care.

DR. PORT: Mammograms do save lives. Those of us who are the first line of defense for our patients with breast cancer see this every day. Mammograms obviously don't *prevent* breast cancer, but they are one of the main reasons for optimism if you do get diagnosed. If you get a yearly mammogram and you do develop breast cancer, that cancer is very likely to be picked up early, when it is significantly more treatable and curable. The science to support this has existed for almost fifty years now. From the earliest mammogram screening trials done in the 1970s and '80s, we learned that women who got regular mammograms had a higher survival rate than women who didn't across *all* age groups that were studied, beginning at age forty. The studies showed that if you were forty to forty-nine and getting yearly mammograms, you were 15 to 20 percent less likely to die of breast cancer, thanks to early detection, compared to women who were not undergoing screening. If you're not undergoing screening, a cancer would not be detected until there was a change on exam by a doctor or you noticed something yourself: a large enough lump, skin tethering, nipple inversion, an enlarged lymph node under the arm, all signs of more advanced cancer. Mammograms have saved lives: it's a no-brainer. Their benefit only increases with older age groups, where breast cancer becomes more common.

One of the questions that I am asked increasingly by my patients (given the aging population) is, "At what age should I stop getting a mammogram?" The benefits of screening mammography have not been widely studied in the over-seventy-five age group, as most screening studies did not include women above that age. That's because screening mammogram studies were done in an era when the average life expectancy wasn't much longer than seventy or seventy-five! Times have changed, and most of us who practice believe that as long as a woman's life expectancy (based on other medical conditions and overall health) is longer than five years, the benefits of screening mammography beyond age seventy-five most likely continue. Many of my eighty-five-year-old patients are playing tennis and golf, socializing, and living their best lives. I know it sounds like a silly rule, but if you can still pick up a tennis racket or a golf club, you are healthy enough to get a mammogram.

The undeniable benefits of mammograms don't make it easy to sift through all of the new information that's out there. Over the past few years, there has been a huge amount of conflicting advice about mammograms: When should you start? How often should you get them? At what age should you stop? For many years, all the major medical societies, along with the American Cancer Society and the government agency charged with developing guidelines for screenings and balancing benefits and harms, the United States Prevention Service Task Force (USPSTF), agreed on one clear, united message and guideline: get a baseline mammogram at age forty and keep getting them annually after that. That should have been the end of discussion, but it wasn't.

Starting in 2009, consistency in the mammography screening guidelines got disrupted when the USPSTF changed their recommendations to advocate *against* routine screening mammography in the forty-to-forty-nine-year-old age group, and to have one every other year after that, in the fifty-to-seventy-four-year-old age group.

Even more confusingly, in 2015, the American Cancer Society also changed their recommendations, but not to align with USPSTF. They completely abandoned the recommendations they had stood by since 1992, now stating that mammograms should be done yearly from ages forty-five to fifty-four and should change to every other year at age fifty-five. Before age forty-five, mammograms can be done selectively. If it sounds confusing, that's because it is. The lack of agreement or consensus among the most respected organizations for mammogram guidelines since 2015 has left the public, and even doctors who were charged with ordering mammograms for their patients, baffled. Which guidelines should they follow? Would the tests be covered by insurance, or would the patients now have to pay out of pocket for following one guideline and not another?

These changes in guidelines got a lot of attention. The question most people asked was "Why?" Why recommend *against* mammograms? Well, it's not because they don't save lives. That part was not disputed by any organization. So why the change? Ultimately, it was due to a shift in prioritization. The leadership of these organizations effectively said, "Hey, let's put cost containment and avoidance of the anxieties and harms related to potential false positive findings at the top of our list and deprioritize saving lives and early detection." The rationale was that less mammography in the younger age groups would lead to fewer biopsies and, in turn, less anxiety for an age group less likely to develop breast cancer. But what about those who did develop breast cancer young? What about the greater anxiety and stress that comes along with a more advanced cancer diagnosis due to lack of screening?

Back in 2015, I was so frustrated by the change in mammography recommendations that I cowrote an opinion piece about it for *The New York Times*, along with two breast radiologists. At the time, the three of us were witnessing on a daily basis the edge that early diagnosis and treatment of early breast cancer provide—and the anguish and devastation of failure to do so. We felt the need to communicate our continued support of yearly mammography beginning at age forty and our continued priority to save lives over the other endpoints, like cost cutting. Boy, were we eviscerated. In the online comments section, most remarks ran along the lines of: "Of course they recommend screening. If the screenings stop or decrease, the radiologists are out of a job, and if breast cancer isn't detected, so is the surgeon!" Scrolling through the endless, grueling criticism was disheartening and insulting, both personally and professionally. It was a rude awakening to the far too commonly held conspiracy theory that doctors promote screening so we can detect problems and line our pockets by treating and curing them. While there are many people out there promoting unproven remedies, treatments, and tests, we are not them, and mammograms are not those. The data and years of research involving hundreds of thousands of women and decades of follow-up are indisputable. We know that mammograms, though not perfect, are effective at detecting cancer earlier, translating into improved survival.

What we don't talk about as much are other secondary benefits of mammography. They may be less important than saving lives, but they can be critical to quality of life. In 2018, my team at Mount Sinai published a study that demonstrated findings about the benefits of screening that hadn't been described until that point. We took our entire population of patients who came in for a new diagnosis of breast cancer and categorized them into three groups: women who had been undergoing regular screening, whose last mammogram was in the prior two years; women who had at least one prior mammogram, more than two years prior; and women who had never had a mammogram ever. Patients in the first group, who had undergone some level of mammographic screening and were more likely to have their cancers detected that way, were significantly more likely to have the option of less extensive surgery, such as a lumpectomy instead of a mastectomy. They were also significantly less likely to need more extensive lymph node surgery due to cancer spread. Finally, patients who were screened were less likely to require more aggressive medical treatment: chemotherapy, which in my experience, patients often dread the most. So, our research findings showed that early detection with mammography is associated not only with improved survival but also less aggressive surgery and treatment.

In 2024 the USPSTF, at long last, issued an update to their recommendations, which I would call a partial reversal. Based on data showing that younger women, especially Black women, are at risk for developing more aggressive forms of breast cancer at younger ages, they reverted back to recommending that mammograms start at age forty. Thankfully, this was a step back in the right direction, supported by long-standing data. The recommendations validated what I saw in my practice every day, but they didn't go far enough. The recommendation for the forty-year-old start was spot-on, *but* they specified every *other* year, not yearly. The impetus for earlier screening was based on new data demonstrating that young women who develop breast cancer are more likely to have aggressive disease and die from it. Aggressive cancers can grow quickly; a lot of progression can occur in two years. This is why yearly screening is more effective as it optimizes the chance of early detection. Who knows how many breast cancer–related deaths have happened because of the flip-flopping recommendations and the confusion they caused.

Last, I would like to offer one more potential insight regarding the big-picture benefits of yearly screening. Most people believe that breast cancer research is focused on newer and improved therapies, more aggressive treatment, and better backup treatments for patients who break through conventional therapy, what I refer to as "longer and stronger." That is true, but it's not the full picture. My colleagues and I, who are in the trenches fighting this disease every day and care deeply for the people we treat, will not be satisfied until the death rate from breast cancer is zero. It's an ambitious goal; we have a long way to go, especially given that the current yearly death rate is approximately 43,000.

When we talk about breast cancer research, we often neglect the fact that a major focus for the past two decades has been figuring out how to customize, tailor, and potentially *dial down* or eliminate more aggressive surgery and treatment for some individuals while delivering the same survival. Here's more good news: we've made real progress in that respect. We've gone from regularly performing radical mastectomies to lumpectomies. From taking all the nodes out under patients' arms to only a few whenever possible. In some cases, now, we may not remove any nodes at all. We've gone from recommending radiation to all women who undergo lumpectomy to defining groups for whom radiation does not add benefit and can therefore be omitted. Chemotherapy used to be routine for most patients with invasive breast cancer, but research led by one of my Mount Sinai colleagues, Dr. Joseph Sparano, a medical oncologist, demonstrated that chemotherapy can now be safely avoided in many patients without compromising survival. We've seen a 30 percent reduction in delivery of chemotherapy to breast cancer patients over the last decade. As I write this, a number of my surgeon colleagues are leading the charge in trials to determine whether surgery can be avoided altogether in cases of early breast cancer, or in cases where the tumor responds completely to other forms of therapy. To put it simply, surgeons are advocating for less extensive surgery. Medical oncologists are leading the effort to dial down the need for chemotherapy. You will read much more about this theme of "de-escalation" of treatment on all fronts in the chapters to follow on breast cancer treatment. But here's the thing: the potential for less treatment across the board can *only* be accomplished with early detection. Right now, there is only one tool proven to reliably accomplish this: the mammogram.

Now, to address the elephant in the room: Are mammograms perfect? Of course not. No test is. When I speak in public about breast cancer screening, the most common questions I get during the Q&A portion go something like this:

My sister was diagnosed with breast cancer when she felt a lump one month after a normal mammogram. Why should I get one if it didn't work for her?

I heard that sonograms and MRIs are better than mammograms at picking up cancers in some women, especially those with dense breasts. I have dense breasts. Why aren't those tests also recommended?

I have heard that mammograms can overdiagnose cancers, find small things that might never have caused a problem for me in the first place! Is that true?

To be clear, mammograms are the best screening test for the *general* population and do pick up 80 to 90 percent of breast cancers for most groups. We do recommend additional tests (which I'll get to in following chapters) on an individualized basis determined by a person's risk. It's also important to know that the technology for different kinds of mammograms has improved. Approved in 2011, 3D mammograms, now offered in approximately 50 percent of screening facilities across the country, are even better than traditional mammograms, and are considered state of the art. They pick up more cancers and generate fewer callbacks and false positive findings (that is, when they find "something," but it ultimately ends up not being cancer, a good outcome, but a stressful experience to go through). This is why you should consider asking for 3D when you go for your yearly mammogram.

There is also contrast-enhanced mammography, where an

iodine dye injection helps highlight cancers on a mammogram, which can pick up more cancers than regular mammograms in women with dense breasts. A recent, extremely important study from the UK showed that for women with dense breasts, the addition of a contrast-enhanced mammogram picked up almost twenty more cancers per thousand patients screened. Contrast-enhanced mammograms are not yet FDA approved for screening and are not as widely available. The iodine dye injection is also associated with a risk of allergic reaction, which is a concern for many. They are also more expensive than regular mammograms, although they are cheaper than an MRI.

Right now, the next frontier of mammograms lies in applications of artificial intelligence. Most mammograms are read by human beings who are doctors called radiologists or breast imagers. As precise or careful as any human being is, we all get tired, miss things, and make mistakes. To enhance the radiology and imaging we already do, hospitals are starting to train and implement AI algorithms to read mammograms and detect cancer. In my health system, Mount Sinai, all mammograms are *double* read by a radiologist and AI, giving us double the chances to catch something early. So, what might sound like a futuristic idea of health care is actually current practice.

Finally, also on the topic of cutting-edge utilization of mammography, research performed by my colleagues at Mount Sinai and others across the country indicates that mammograms may not only be important for breast health but might surprisingly play a role in "heart health" as well. How can a mammogram tell you anything about your heart? Well, mammograms can see blood vessels that run through the breast. If the breast blood vessels, specifically the arteries, look white or calcified, this could be a marker for vascular disease and a first hint of similar disease elsewhere in the body, such as the heart, even when a woman has no other symptoms. Previously, breast arterial calcifications (BAC) were ignored. Now, many breast radiologists advocate for identifying and reporting increased vascular calcifications in the breast and informing women of these findings via their mammogram reports. These results can then be discussed with one's doctor, who can then initiate evaluation of the heart and other vascular disease. One of our Mount Sinai patients was recently informed of her BAC, saw a cardiologist, and was found to have severe coronary artery disease that she did not suspect. She underwent urgent open heart surgery, and credits her mammogram (and the breast radiologist who read it) for saving her life.

Now, people who've gone through the roller coaster of a false positive, or know someone who has, may be wondering: Can mammograms *overdiagnose* cancer? This is a very controversial topic and relates to so many cancers. Overdiagnosis and overtreatment mean something is discovered and then treated, but the person might have done fine without that diagnosis or treatment. Of course, overdiagnosis and overtreatment issues are much more relevant in older populations, who arguably have shorter lifespans. There is so much research now on "watching and waiting" on an actual cancer diagnosis and figuring out who are good candidates for this approach, which is certainly not a one-size-fits-all. In general, most breast cancers, if left alone and untreated, will progress over some period of time, and most cancers picked up on mammograms would go on to cause harm if left untreated. More on this later in the chapter on DCIS, the earliest form of breast cancer, where the watchful waiting approach is increasingly being considered.

6

MORE SCREENING OPTIONS

WHERE DO ULTRASOUNDS OR SONOGRAMS FIT IN?

An ultrasound is also called a sonogram; they are the same. This is a type of radiologic test that does not use radiation, but rather sound waves. It can see, clarify, and localize abnormalities that we already see on a mammogram, but it can also identify new findings, such as lumps, that mammograms don't see.

VANESSA H., tollbooth attendant: I had been getting yearly mammograms since I was forty. Now, I'm fifty-two. I was doing everything I was told to do. I have no real family history of any cancer, other than my mother who got lung cancer from smoking, which I don't do, so I assumed that the "basics" were good enough for me. A few years ago, my mammogram report came in the mail, as usual, but it wasn't the same as the ones I'd always gotten. This time, it told me that my breasts are "dense." I had no idea what that meant or what the significance of having dense breasts was. The report went on to say that dense breast tissue can make it harder to see a cancer on mammograms and that I should discuss "additional imaging" beyond mammography with my doctor.

I called her and she said that I was only finding out about my dense breasts now because of a new notification law in our state that mandated disclosure of breast density after mammograms. My doctor went on to say that based on this information, we could decide together if I should get an ultrasound, which can pick up some cancers that mammograms miss for women with dense breasts. She warned me that they can also pick up a lot of background noise. False positives, she called them, which, in turn, can lead to callbacks, biopsies, and even surgery, if there is uncertainty about what the sonogram shows. "The more you look, the more you find," my doctor explained. She also advised that I go ahead with the ultrasound, as long as I knew what might happen. I had the ultrasound. Thankfully it was clear, and I now have a sonogram exam every year with my mammogram, which gives me added reassurance that nothing will be missed.

DR. PORT: As we discussed, mammograms save lives, but no test is perfect. Mammograms pick up 80 to 90 percent of breast cancers, but there are factors that increase and decrease the likelihood of mammograms picking up a cancer. The main one is having dense breasts like Vanessa.

So, what *is* dense tissue? Well, the breasts are composed of two types of tissue: fat and fibroglandular tissue. The fat part is easy. We all have it in varying quantities and on a mammogram, you can easily see through fat. Fibroglandular tissue is the functioning tissue of the breast, the stuff that produces milk and delivers it to any potential babies that might come along. Naturally, fibroglandular tissue predominates in young women of childbearing age. Up to 50 percent of women in their forties have "dense breasts," meaning less fat and more fibroglandular tissue. When women in their forties come to me saying they're concerned about being told that they have dense breasts, I usually tell them not to worry. Half the women their age do too. As women get older and breasts no longer need the baby-feeding function, a natural part of aging is the atrophy of dense, fibroglandular breast tissue. Breasts tend to become fattier, which facilitates breast cancer detection, a plus since cancer risk increases with age. However, some women do retain significant breast density, even as they age, which makes it more difficult to detect cancer. And having continued breast density into the older age groups does increase the risk of breast cancer to a small degree.

Why does dense breast tissue make it harder to detect breast cancer? Denser breasts look whiter on a mammogram. Cancer also shows up as white. So, a white cancer against a white background can make it harder to see, the proverbial polar bear in a snowstorm. Radiologists rate breast density on a four-level scale:

A. fatty, meaning not dense at all

B. scattered density

C. heterogeneously dense, or mixed density with a higher proportion of dense tissue

D. extremely dense

For women in categories A and B, with more fat than dense tissue, the sensitivity of mammograms is quite good, 90 percent or higher, according to some studies. However, for women with dense breasts, C and D, that sensitivity can go down to 60 or 50 percent, so the chances of missing a cancer on a mammogram are significantly higher. How do we close this gap and make sure that women with dense breasts at *any* age have the highest chances of early detection? We add other tests, usually starting with an ultrasound.

Ultrasounds can pick up some of what mammograms miss, and are an added value, especially in women with dense breasts. Which brings us back to Vanessa's story, and the new breast density notification on her mammogram report. In 2004, Nancy Capello was diagnosed with advanced stage breast cancer despite years of normal mammograms, after feeling a lump shortly after yet another normal mammogram. When she went back to her doctor, an ultrasound of the lump that she'd felt visualized a large mass that ultimately proved to be a cancer. She became outraged when the explanation given as to why the mammogram missed the cancer was that her breasts were dense, which she knew nothing about. Her conclusion was that if she had actually known about her breast density, she might have asked for additional imaging, which may have detected her cancer earlier. She and her husband developed a nonprofit called Are You Dense? to raise awareness about screening for breast cancer with dense breasts. Their goal was to create legislation to mandate patient notification on their mammogram report regarding their own breast density. Furthermore, they wanted to make sure that insurance companies would cover the expense of that ultrasound. In 2009, Connecticut passed the first law requiring doctors to notify their patients of breast density, empowering patients like Vanessa to at least have a conversation about additional imaging. By 2019, thirty-seven additional states had followed suit. Finally, in 2024, the FDA supplanted state legislation, requiring *all* accredited mammography facilities across the country to notify patients of their breast density as a part of the mammogram report that all women receive after their exams.

Getting a letter in the mail that says you have dense breasts may cause anxiety, but one of the mottos of this book is "knowledge is power." We should never shy away from information that can have health implications. It's good to know, and you can always decide not to do anything differently. Informing women of their breast density is a piece of knowledge that can save lives in a small percentage of patients. But there's no doubt that this new law created confusion among patients, primary care providers, gynecologists, and many others. "What should I do?" many patients asked, turning to my office for advice from breast experts. "How should I advise patients?" my primary care colleagues would call to inquire. I'd tell them that there is no one-size-fits-all solution, but if you have dense breast tissue, and especially tissue that remains dense as you get older, you should definitely consider adding an ultrasound to your yearly mammogram.

Here's the caveat: it's important not to *overestimate* the benefits of sonography or to think that it can replace mammography as the sole screening test for breast cancer. An ultrasound is usually done by an experienced technician who "scans" the breast and goes segment by segment looking for abnormalities. However, an ultrasound is never a *complete* picture of the breast the way a mammogram is. Ultrasounds can miss spots too.

Take my patient, Robin, who came in to see me after her ob-gyn felt a small lump in her left breast one week after a normal screening mammogram *and* ultrasound. I felt the lump, too, and sent Robin to our radiology suite for a *targeted ultrasound,* which is totally different from a screening ultrasound. With a screening ultrasound we are looking everywhere, scanning both breasts in a very general way. With a targeted ultrasound, we are looking more closely in a very specific area where we have already identified something that we are concerned about; either something that we feel, or something that we already see on a mammogram. I marked what I felt on Robin's skin with my purple marker, and told my breast ultrasound tech, "Look there." She saw a suspicious mass directly under the spot that I had marked and performed a biopsy of the lump, which proved to be cancer. So, what happened the first time around just a week earlier? Was the first exam not thorough enough? How did they miss the spot on both the mammogram *and* the sonogram? The answer is that yes, sonograms can miss spots, even in the best of hands. That's why it should not replace the mammogram and is not the best screening test *by itself.* Robin has told the story as the sonogram finding a cancer that the mammogram missed. But the truth is that her ob-gyn found the cancer. The ultrasound identified it only after we told them where to look. This distinction is important: the ultrasound *seeing* the cancer after we told them where to look is not the same thing as an ultrasound picking up a cancer via a routine screening exam. This story also underscores the importance of being specific when telling our health stories, so others don't get the wrong message.

MRIs

FOR SCREENING HIGH-RISK PATIENTS,
WHEN DIAGNOSED WITH BREAST
CANCER, AND FOR FOLLOW-UP

MEREDITH D., hotel receptionist: I knew I was at a high risk for breast cancer. My two sisters both had it. Both of their genetic testing was negative, but we knew there was something going on in our family. Maybe it was a gene that we just couldn't test for yet or one that hadn't been discovered. Whatever it was, I knew I had to be diligent; I got my mammogram and sonogram every year. One year, when my gynecologist felt something in my left breast after my mammogram and sonogram were normal, she sent me to see Dr. Port, who had taken care of both of my sisters. Thankfully, Dr. Port didn't feel anything on my exam. She told me that what my gynecologist had been concerned about was just normal dense breast tissue. After a long discussion about how I should be screened, given my high risk, we talked about adding on MRI, a test, she explained, that is not usually recommended for screening the general population. It is recommended, however, for women, like me, who qualify based on their risk level. MRI can pick up cancers that mammograms and sonograms can miss, so when you are high risk like I am, you want to do everything you can to not miss a cancer. The MRI came back and showed a spot in my right breast that needed a biopsy. Guess what: the biopsy showed the earliest possible stage of cancer. I couldn't believe it. I had walked into the doctor's office concerned about one thing that my gynecologist wanted me to have checked out and found out it was nothing, only to learn about something else that turned out to be a real, actual cancer. I kept wondering what would have happened if I hadn't gotten that MRI. As Dr. Port explained, it is possible that the area would have been identified on the next round of imaging, but who knows, and the next round wasn't for another year. By then, it could have been more advanced. Thankfully, as a result of having an MRI, my cancer was caught early, at stage 0. If I was going to get breast cancer—and based on my family history, I figured I probably would— this was the best-case scenario.

DR. PORT: Magnetic resonance imaging (MRI) for breast cancer was developed in the 1980s and '90s, but it became more widespread and commonly utilized in breast cancer screening after 2000. While MRI is the most sensitive test for picking up breast cancer, it is not done *instead* of mammograms; rather, it's an additional screening measure for appropriately selected individuals. Believe it or not, no matter how good an MRI is, mammograms are still better at some things, like picking up calcifications, which can be a sign of early breast cancer. A breast MRI—which is not the same as an MRI of other body parts—can be a more difficult test for a woman to have. Whereas snapping a mammogram picture takes mere seconds, a breast MRI takes longer, usually about thirty to forty-five minutes to complete the test. MRIs involve having both breasts scanned and an IV placed with an injection of a type of contrast dye containing gadolinium, which helps light up or highlight areas of increased blood flow, which includes cancers. It does not involve radiation, but there is a chance the dye may come with a potential risk. Currently, no data clearly show that it has harmful effects, but there's been some concern that it may accumulate in the brain and that repeated MRIs over years and years may have adverse effects. It's important to weigh both factors with your doctor as you opt either for or against having MRIs as follow-up in the long run.

MRI is typically offered or recommended in one of three scenarios:

1. To screen high-risk individuals, or, less commonly, women with extremely dense breasts
2. After a breast cancer diagnosis, to see if there are other areas of concern in either breast
3. As follow-up in women who have had breast cancer

Let's take a deeper dive into each of these scenarios to better understand how we, the doctors, decide how to use MRI.

If an MRI can pick up almost everything, why *don't* we recommend them for everyone? The answer is that for the population of patients who most commonly get breast cancer, mammograms are better, all things considered. Now, to understand what that means, we have to define what makes a good screening test. There are a lot of considerations, starting with the sensitivity in picking up cancer. But we also consider another metric called specificity. Specificity is determined by how often a test finds "something" that ends up *not* being cancer—a false alarm, or as we call it, a false positive, which, as we already discussed, is not a good thing. MRI is associated with a high degree of false positives, which lead to additional imaging studies, biopsies, and all the anxiety associated with a looming potential diagnosis. So, when we think about screening the general population, the risk of generating false positives is thought to outweigh the very small chance of finding a cancer, about 1 percent, for someone who is not at high risk. Of course, there are exceptions. Women with extremely dense breasts may be recommended to undergo MRI screening, even if they have no other risk factors, since their mammograms' ability to detect any cancer might be extremely compromised.

SCREENING FOR HIGH-RISK PATIENTS

We do recommend MRI screening for those at high enough risk for developing breast cancer, particularly those at risk for developing it at a young age, in their thirties and forties, when breasts are normally dense. There's more to come in later chapters on the details of how to figure out if you qualify for MRI screening, but, as a rule, if you carry a genetic mutation putting you at extremely high risk of getting breast cancer, MRI screening is mandatory. If your lifetime risk of getting breast cancer is 20 percent or greater, MRI screening should at least be a discussion with your doctor.

AFTER A DIAGNOSIS OF BREAST CANCER

When a woman is diagnosed with breast cancer on "conventional imaging," MRI can play an important role for some women in determining whether or not there are other areas of concern in either breast. Even though many women get an MRI when diagnosed with breast cancer, MRI is a very polarizing issue in the breast cancer treatment world. Some doctors think that *every* woman diagnosed with breast cancer should have an MRI; others would say that very few need one. Most of us are somewhere in between. I believe that there is no one-size-fits-all solution; each decision should be made on a case-by-case basis. When the findings on mammogram or sonogram are vague, an MRI can be extremely helpful for me as the surgeon to define the extent of disease and aid in the planning of surgery.

Just like in the screening setting, MRIs can be overly sensitive and identify areas of "concern" that prove to be nothing. To get into the statistics, an MRI will identify other areas of concern in women with breast cancer approximately 30 percent of the time. Of that 30 percent, only a small fraction will be actual cancer. Critics of the "MRI-for-all" approach point to the fact that when an MRI demonstrates other findings in a woman diagnosed with breast cancer, her surgery or treatment is often unnecessarily delayed by additional testing and biopsies that turn out to be nothing. Critics also note that MRI may overdiagnose, finding other tiny areas of cancer in the breast that don't need surgery but might need other treatments. Overdiagnosis can confuse decision-making. Remember, we have been doing lumpectomies successfully since the 1980s, long *before* MRIs were available, and recurrence rates are also low in those who don't have MRIs. More likely than not, the additional treatments, like radiation and chemotherapy, that women get after their breast cancer surgery "mop up" microscopic disease that might be left behind, the kind of disease that MRI often shows us. In general, many women undergo MRI once diagnosed with breast cancer and the results can definitely impact plans for surgery and our decision-making for how to advise our patients.

MRI FOR LONG-TERM FOLLOW-UP

Perhaps the most controversial and least defined aspect of MRI utilization is determining who should have them for yearly follow-up screening after breast cancer treatment. When someone has had a lumpectomy (where a small part of the breast is removed) or a mastectomy (removal of the whole breast) but only on one side, the risk of recurrence of breast cancer or development of a new one on the other side is generally quite low. MRIs are often used in follow-up for women whose cancer was not seen well on mammograms in the first place. For women with dense breasts, detecting a recurrence may be more challenging. That's why we frequently offer MRIs as follow-up to women diagnosed with cancer before the age of fifty.

MRIS AREN'T FOR EVERYONE

There are some people who can't get an MRI. Since the test uses magnetic fields, individuals with some types of metallic body implants may not be able to have them, for example women with pacemakers. This does not apply to joint replacements made of titanium, the most common material, as they don't interact with magnetic fields. Others may have a specific allergy to gadolinium, although it is extremely rare and affects only approximately 0.1 percent of the population. Because it also involves lying flat, facing down, and entering a tubelike scanning machine, claustrophobic women may have a hard time completing the test, feeling anxious from the sense of being enclosed for this amount of time. Unfortunately, no alternatives or high-quality "open MRIs" for breast screening currently exist. While we can offer antianxiety medications to patients who really need the MRI, some patients simply can't tolerate the test.

MRI is one of the technologies that has truly advanced breast cancer screening, treatment, and follow-up care in the last twenty-five years. As our technology options increase, there is a lot to learn—as a patient. Each person, with guidance from her doctors, must decide which tests she should have or not have depending on her own specific circumstances and whether or not she fits into one of the specific groups for which MRI has been shown to be beneficial. As an aside, it's important to know that, as with all breast imaging tests, MRIs are best when done in centers of breast specialization. Radiologists look at these images and interpret them, scanning for areas of "enhancement"; where the dye goes indicates increased blood flow, which cancers usually have. Having an expertise in breast *imaging* as a specialty means your radiologist has a dedicated focus and probably more experience in "reading" breast MRIs, which then increases the chance of not missing something important, while also not over-calling findings that could lead to unnecessary procedures. Will AI also be incorporated into MRI reading and interpretation as we're seeing with mammograms? Stay tuned. Looking at medical imaging studies of all types and recognizing abnormalities or change from prior studies is one area where AI is rapidly proving to be advantageous in our field of medicine.

8

A BIT ABOUT BIOPSIES, AND SOME OF THE *NORMAL* THINGS WE MIGHT FIND

SUZANNE R., retail store manager: I had been getting my mammogram every year, and then an ultrasound was added. I don't find the mammogram to be pleasant, but then neither is a colonoscopy, or a Pap smear, or any of the other screening tests that I go for. Even my skin check involves me standing basically bare ass naked in the doctor's office while she looks me over everywhere in every nook and cranny, if you know what I mean.

The point is: if you do all of these screening tests, which give you the best possible chance to pick up cancer—*any* cancer—early, you are bound to find something at some point: a skin lesion on my back needed to be taken off. I had a colon polyp taken out. So, I tried to keep it in perspective when I was told a couple of years ago that I had a new mass seen on my mammogram and then the sonogram in follow-up. I ended up with a needle biopsy, which showed a fibroadenoma, a completely normal mass that, as long as it doesn't grow or change, which it usually does not, can be left alone. So, like my other screening tests and biopsies, which showed stuff, this was a false alarm, but a necessary thing to do to make sure I didn't have a new cancer. The process is stressful, but I was so relieved. I'm going to continue to get checked, so that if one day it's not normal, I have the best chance of early detection.

DR. PORT: *Biopsies.* Let's talk about part of the screening process that people worry about: when something comes up that requires a second look. This usually means first getting called back in for additional pictures, which in the majority of cases will reveal nothing of concern. Phew! If that's not the case, the next step is getting a biopsy. Biopsies are done with a thick needle called a core needle, which takes out snippets of tissue from the area of concern. These biopsies can be done by a surgeon or more commonly by a breast radiologist who uses one of the imaging studies you just learned about to guide them to the area of interest. The type of imaging that the radiologist uses to guide them depends on which imaging test provides them with the best view of the target. So, if a radiologist is doing a biopsy of calcifications, small white speckles that are commonly seen only on mammograms, she'll put the patient in the mammogram machine and use the images as a guide to the area of interest. Similarly, when the area of interest is only seen on an MRI, the person having the biopsy goes back into the MRI machine and has the biopsy that way. When a biopsy is guided by a mammogram, it's called a stereotactic biopsy. When the biopsy is guided by an ultrasound, it's called a sonogram-guided biopsy. When guided by MRI it's called an MRI-guided biopsy.

There is a big myth that needle biopsies can spread cancer. Let me be clear: it is not true, and I wish this bit of misinformation would just go away! Regardless of the imaging technique the radiologist uses, the needle is essentially the same size, and so is the amount of tissue removed. These small samples of tissue go to the pathology laboratory, where a pathologist, ideally one specialized in breast pathology, processes them and looks at them under the microscope. Usually, based on these specimens of tissue, we can tell if the area is "benign" (normal); inconclusive, which we'll discuss further in later chapters; or "malignant" (cancer). Needle biopsies are the standard of care for figuring out what something actually is, and then we will know how to deal with it.

LOCALIZATION

A needle biopsy may show something that will need to be removed with subsequent surgery. This is true of course if we find cancer, but we do remove other things as well, including some benign tumors that should be taken out, or areas for which we need more information that isn't available with a simple biopsy. There are many things found through imaging that we surgeons can't feel with our hands or see through the skin. We find and surgically remove a small area that we are looking for with the help of a technique called localization. At the time of the needle biopsy the person performing it leaves a small marker or tag in the area where the biopsy was performed. This "X-marks-the-spot" approach ensures that we can find the precise area if we need to remove it when the biopsy results come in.

A localization involves the placement of a wire or another different tag that gives off a signal into the breast. A wire is usually placed by a radiologist right before the surgery. Using local anesthesia, the radiologist advances the wire into the breast as a form of guidance to the area of interest. In surgery, the surgeon makes an incision and follows the wire down to the area of interest and removes the area in and around the wire. Alternatively, a tag or seed can be placed into the breast days before surgery, next to the biopsy tag, and it emits a signal. Once in surgery, the surgeon scans the breast with a probe that picks up the signal from the tag. Again, we make an incision and follow the signal down to the area of interest. Localization is part of any breast surgery that involves removing an area of the breast, whether cancerous or normal, that we surgeons can't readily feel or see. It's a way of making surgery more precise and directing us to give us the best possible chance of getting the correct area out. Importantly, if a biopsy result is normal, and nothing needs to be removed, the tag is harmless and inert and can stay in the breast forever. It won't set off metal detectors in the airport, and rest assured, you won't be able to feel it.

This process of localization is an important part of many breast surgery procedures and is critical to understand for those who might need to have something removed. Thanks to the use of local anesthesia, it is usually painless.

BIOPSY RESULTS ARE USUALLY BENIGN

Many women of all ages develop lumps or bumps in their breasts or things are found on their imaging. It's worth reiterating that *all* new breast findings that a woman or her doctor notices should undergo further investigation, regardless of her age. Once imaging is done, often a biopsy, as described above, is needed. Thankfully the majority of biopsy results will be benign (noncancerous). Managing noncancerous findings is an important part of breast health too. So let's go over the most common benign breast conditions and what to do about them.

FIBROADENOMA

The most common cause of a breast mass in a young woman is called a fibroadenoma. These are benign lumps that a young woman typically feels and then brings to the attention of her doctor. They can also be found on imaging like Suzanne's was, appearing like a mass. They are usually very round and mobile within the breast and can feel like a small or large marble. If a young woman feels a new mass, often an ultrasound will be done, and then a needle biopsy. Once a biopsy has shown that the mass is a fibroadenoma, there are two options: monitor the growth by doing another ultrasound, usually six months later, or take it out. We typically remove them when they are large to begin with, greater than two or three centimeters (more than an inch); when they have shown growth on follow-up; or out of patient concern or desire. It's also advisable to remove them in a young woman who is considering getting pregnant in the near future. Fibroadenomas can grow with the hormonal stimulation of pregnancy, and the last thing a woman, or her doctors, wants to worry about during pregnancy is an enlarging breast mass and performing surgery on a pregnant woman.

PASH

PASH (pseudoangiomatous stromal hyperplasia) is another source of benign breast lumps. If a biopsy is performed showing this, the area typically does not need to be removed. PASH involves cells of the breast's connective tissue, called the stroma, that show overgrowth and can therefore feel like a lump when clumped together. The area usually does not grow or change and only needs to be removed if there is concern of something else other than PASH, or if the area is painful or bothersome.

SCLEROSING ADENOSIS

Sclerosing adenosis is also benign and a potential cause of feeling a lump in one's breast that does not normally need to be removed.

When a biopsy shows normal findings such as those above, we are of course relieved. We, surgeons, often review everything and ensure that we conclusively believe the results. If something feels particularly concerning or looks highly suspicious on imaging, but the result is normal, this is termed a *discordant biopsy*: the results and the appearance don't match up. In such cases, often the safest thing to do is a small surgery to take the area out and make sure nothing was missed on the biopsy.

OTHER BENIGN THINGS WE OFTEN REMOVE

Radial scars are benign, but again are sometimes removed because their star-shaped irregular appearance on imaging can make them look more concerning.

Papillomas are benign but complex lesions that can form within the breast, particularly in the milk ducts of the breast directly behind the nipple. Interestingly, papillomas are the most common cause of nipple discharge, either clear or bloody. Often they need to be removed to definitively rule out cancer. And occasionally, a small cancer might be found living within a papilloma.

Phyllodes tumors are rare and strange masses that are most often benign. We remove them as they commonly get quite large. Differentiating fibroadenomas, above, from phyllodes tumors can be challenging, and sometimes a biopsy report will say that with the small sample given, the pathologist cannot definitively determine which one of these it is: a fibroadenoma that can be left alone or a phyllodes tumor that should be removed. The language often used in the biopsy report is "likely fibroadenoma, but cannot rule out phyllodes tumor." While most of these often do end up being just fibroadenomas, we frequently take them out just in case. There are rare cases of malignant phyllodes tumors that can be quite concerning and often are very large due to their rapid growth. These require complete removal and may even require mastectomy depending on the size.

9

CUTTING-EDGE SCREENING TECHNOLOGY

FULL-BODY SCANS, LIQUID BIOPSIES, THERMOGRAPHY—WHAT'S READY FOR PRIME TIME AND WHAT ISN'T?

AMELIA J., corporate lawyer: I'm in my fifties, and, like so many friends my age, I find it hard to tune out the noise of medical advice online. There are so many people out there promoting medical tests: full-body scans, thermography for breast cancer, the list goes on. Some of them cost thousands of dollars, but the implication is that if you can afford it, you should do it. Why wouldn't you want to know if you have cancer? Why wouldn't you want to pick it up early, if you can? I learned the hard way that these tests can open up a Pandora's box. I signed up for and did the test that everyone seemed to be talking about: a full-body scan. It showed something in my belly, maybe in my pancreas. It wasn't clear if it was a cyst, a benign tumor, or something worse. When I called my doctor in a panic about the results, he was at a loss because I'd signed up for the test without telling him. He consulted with other doctors, who recommended a dedicated CT scan of the abdomen. I did it, and it showed a small mass in my pancreas that I was told was most likely benign. I thought that was ridiculous. "Most *likely* benign? Isn't pancreatic cancer, like, one of the most lethal cancers you can have?" I asked my doctor. I needed to know if it was *definitely* benign.

"Well, the only way to do that is to have a biopsy, which is pretty invasive, or to take it out in an operation, which is very invasive," my doctor said. "Both of these have risks. Our recommendation is to do a follow-up scan in three months. If it hasn't changed, we should be fine."

"And if it has?" I asked, panicked.

"We'll cross that bridge when we get to it, but yes, you would probably need one of those invasive procedures that I mentioned."

"Those are my options? Nothing else?"

"That's it," he said. He told me it had probably been there all along, I just didn't know before the scan, and that I shouldn't spend the next three months worrying.

Needless to say, it was a rough three months.

At night, when it got quiet, I thought about it a lot. I woke my husband with my intrusive thoughts. Did I have pancreatic cancer growing in me now? Would it be found to be widespread on my next scan? I started to get stomachaches. Were they related to the mass? Stress? Something else? Three months later, when I got my next scan, the small mass was exactly the same.

My advice is: don't do the full-body scan unless you

are fully prepared to go on a journey of ups and downs, filled with anxiety and potential long-term uncertainty. The peace of mind I went in seeking got replaced by months of stress when something was found. It was a terrible experience that took three months of my life away. The crazy part is I actually paid for it.

OLIVIA A., wife and mother: I think that a full-body scan saved my life. I went in to have it, certain that the results would be fine. But they weren't. Something showed up on my lung. I was fifty-five years old. Since my mother and sister both had breast cancer, I know I'm high risk, so I see Dr. Port for regular screening. I have never smoked, but my parents both smoked, and I was constantly around smokers growing up: in our house, in the car, even on airplanes. We never actually sat *in* the smoking section, but when my parents wanted to smoke, they would stand at the back of the plane, a few rows behind our seats, and light up. It was the '70s.

I did worry that I might get lung cancer because of all that exposure, but I didn't think too much of it until the scan showed a small nodule in my left lung. I had no symptoms and was told it was probably nothing. Thankfully, my doctor ordered a more detailed CT scan of my chest and lungs, which picked up the mass. After a biopsy, we learned that it wasn't nothing; it was lung cancer. When I saw a lung surgeon, she said, "I would never tell a person diagnosed with lung cancer that she's lucky, but you, my dear, are lucky." I had stage I lung cancer. The cure rate is 90 percent. The cure rate for stage III, when most lung cancer is found based on symptoms, is around 15 percent. I prefer my odds.

Most lung cancer is already late-stage when it's found. By the time you have a cough or are short of breath, the cancer is usually too far gone to cure. I know that scans can generate a lot of unnecessary follow-up tests, but for me, those follow-up tests were absolutely necessary. They saved my life, and I'm grateful for them every single day.

DR. PORT: Patients come in every day to ask me about general health issues, especially years after their diagnosis, when they are presumably cured and healthy. When you have taken care of women for twenty-five years like I have, you become knowledgeable about all kinds of health issues from stories and experiences your patients share. Many people, especially those who have had cancer and benefited from early detection, want to be as proactive as possible about screening for other cancers.

Full-body scans, "liquid biopsies," and other new technologies for cancer screening have exploded in the last decade. There will continue to be a number of other screening tests touted as the next best thing by the time you're reading this. New technology is the way of the future; just look at how mammograms have gone from film screen to digital to 3D and contrast-enhanced. We are also doing much of our newer testing with less exposure to radiation, and using other, newer scanning technologies like MRI, confocal microscopy—a technique that involves images produced by a microscope often using fluorescent dyes to identify abnormalities—and even blood tests to name a few. Don't get seduced by the bright and shiny new screening test that some celebrity is pushing and getting paid handsomely to promote. Remember, they can afford it more easily than the average person. Don't be tempted to spend thousands of dollars on an unvalidated, unproven technology to make yourself feel better that you don't have cancer. Yes, as with Olivia's case, the test might find something important, but as Amelia said, if you choose to do the screening, be prepared to go on a journey that doesn't always have an immediate or even satisfying resolution. Also, if you do choose to get screened in this way, it's worth it to have an *up-front* discussion with your doctor so they can be prepared to guide you through the process if something is found.

FULL-BODY SCANS

There are various types of full-body scans, using various types of screening technologies, but as the name suggests, these are procedures designed to take images of your entire body. Here are the facts: if you have a full-body scan, there is an approximately 10 to 25 percent chance that there will be findings that require further investigation, depending on the type of scan you choose. Less than 2 percent of those findings will prove to be an actual cancer. To be sure, if you end up in that small percentage like Olivia, a full-body scan can be lifesaving. But there is a significantly greater chance that you will end up in a situation like Amelia, pursuing invasive and ultimately unnecessary follow-up tests.

I worry that as more people undertake full-body scans to monitor their general health, they will ask themselves why they need to do a mammogram, a colonoscopy, or a lung CT when the whole-body scan sees it all? In reality, full-body scans do not provide as much detail as scans tailored to specific body parts.

A mammogram or dedicated breast MRI will find significantly more and earlier cancer than a full-body scan. The screening exams that doctors recommend—mammograms, colonoscopies, pap smears, chest CT for smokers—have years of data proving their value in early detection of cancer. If Olivia had talked to her doctors about lung cancer being a concern, she could have done a screening CT of the chest every year, which is often recommended for those with a history of smoking or concern of significant smoke exposure.

Another consideration for embarking on a screening journey is radiation exposure. Some full-body scans involve significant exposures to radiation, while others don't. Ironically, even if the scans don't use radiation, like an MRI or ultrasound, if we find something and need further testing, many of *those* scans use radiation, as do the litany of follow-ups usually needed.

LIQUID BIOPSIES

Liquid biopsies typically take a small sample of blood or other bodily fluid and search within that sample for tumor markers, circulating tumor DNA (also called ctDNA), or circulating tumor cells (also called CTCs). The benefit of the liquid biopsy concept is that it is a relatively noninvasive test and it's easily accessible. The bad news is that these tests are not ready for prime time with regard to screening for many cancers. Remember, there are false positives *and* false negatives. Having a normal blood test does not mean there is no cancer anywhere in one's body. And finding abnormalities in the blood often leaves us with information we don't know what to do with. In one study evaluating Galleri, a blood test promoted for cancer screening, 34 patients

(0.5 percent) out of more than 6,600 patients over the age of fifty were found to have a "cancer signal" but no cancer ultimately diagnosed (false positive). Only 25 (0.3 percent) had a signal with subsequent detection of a cancer (true positive).

However, more recent important data related to the Galleri test has become available. When I first started writing this book, a promising large scale National Health Service study in the UK was underway to determine whether the Galleri test, which was designed to screen for over fifty types of cancer, can prevent late stage cancer detection by testing blood at various intervals over the course of years. Over 140,000 patients between the ages of fifty and seventy-seven were enrolled. In February 2026, results were reported: the study showed no significant reduction in late-stage cancer diagnosis, and the primary endpoint (to validate this test for large scale cancer screening) was not met. The test did find more cancers overall. This was a disappointing result; a setback that clearly illustrates that while a validated blood-based cancer screening test would be amazing—very little risk to patients with samples easily obtained—it has to work. And it has to have added value over current standard screening practices, which this important and very large study did not demonstrate. More detailed data will come out regarding whether there was any benefit to the test related to specific cancer types.

Liquid biopsies are becoming more and more common to track how patients with *known* cancer respond to treatment. The results can potentially allow us to adjust therapies accordingly. While this is still an emerging area, it's exciting and promising. For example, ctDNA (circulating tumor DNA), mentioned earlier, checks for fragments of tumor cells—their DNA—in the bloodstream and has been shown to be very helpful in the care of patients with advanced colon and other cancers. A negative test might mean a low risk of cancer coming back, while a positive test might be a sign of something worse to come. It's also important to know that preliminary studies show that liquid biopsies may not be so accurate for some cancers, like breast, that do not typically shed tumor cells into the bloodstream until later stages. So certainly don't give up your mammogram for this type of test.

It's worth noting that liquid biopsies are not the same as tumor marker blood tests, which have been around for a long time. The PSA test for prostate cancer screening, for example, is widely used and extremely helpful for prostate cancer screening or detection of recurrence in those who had the disease. Lots of men have their PSA levels "checked" as part of a routine physical to screen for prostate cancer. For other tumor types like colon, breast, and ovarian, tumor marker tests are not as accurate and are used less and less because they are not considered reliable for screening. Liquid biopsies for cancer detection are moving forward at warp speed, and I'm convinced they *will* be ready for prime time across many cancer types in the near future. But as always, discuss your concerns and your intentions with your doctor before getting started with liquid biopsy testing.

THERMOGRAPHY

Thermography is a breast cancer screening test based on the premise that cancers generate more heat than normal cells, ergo *thermo* in the name. If you spend a lot of time getting information about breast cancer online via social media groups and other unmediated online forums, you will likely have heard of thermography, as there is a large cohort of people out there who are anti-mammography and anti-radiation, but who tout this kind of test as a good alternative to traditional forms of breast cancer screening. Thermography creates a kind of heat map of the breasts to detect variations in surface temperature and hot spots and, ostensibly, shows us areas of concern. The problem is that this test or technology has never been validated for the detection of early breast cancer in the same way that mammography has. As a result, it is not FDA approved for screening, nor as a substitute for mammography.

By now you have probably picked up the terms *false positive* and *false negative*, which we use to describe a screening tool's efficacy. Thermography gives us both high false positive and high false negative rates. It has a poorer performance in detecting early cancers compared to mammography (false negatives), and it may not measure up for deeper breast lesions, where skin surface temperature changes may be undetectable. At the same time, other, noncancerous processes like infections can generate "heat" from the breasts, leading to a high false positive rate. Unfortunately, the poor performance as a breast cancer screening tool does not stop some doctors from buying a machine, promoting its use, and charging money for the test that will probably not be reimbursed by insurance, since it hasn't been FDA approved for screening. Furthermore, without approval for screening, there can't be set standards for the machine or the technology, meaning that if you get thermography, the quality of that particular test, or the reading of it, may be quite variable, whereas you can trust consistent quality with FDA-approved and regulated screenings. Ultimately, there is no technology to do a biopsy based on a thermography result, which means that if the thermography shows a concerning area, but other imaging tests are normal, you are left hanging with no resolution.

I have often been trolled or roasted on social media after doing interviews advocating for mammography and pleading with women to not forgo this tried-and-true screening test for its more expensive, unvalidated counterpart, thermography. When I did the *Lipstick on the Rim* podcast in early 2025, and we talked about all of the reasons why thermography is not a legitimate way to screen for breast cancer in lieu of mammography, I was happy to find that most of the comments were nice. People thanked me and the hosts, Molly Sims and Emese Gormley, for the clear explanation and reliable information. One person wrote, "Thank you so much for this episode and addressing some of the dangerous misinformation about breast cancer (i.e., thermography as an acceptable screening alternative)." She went on to say, "As a recent survivor of an aggressive breast (cancer), it makes me so sad/mad to see women falling victim to predatory wellness." I could not have said it better myself.

The episode also generated a lot of negative comments. The sheer volume of skepticism, paranoia, and flat-out conspiracy theories was mind-blowing. "She is so full of shit," one commenter wrote. Another chimed in, "What does she know?" After taking care of tens of thousands of patients and saving almost the same number of lives over the course of twenty-five years, I didn't think I would need to justify my opinions in the field that is my entire life's work. The sad thing is, it's not totally the commenters' faults; this is our information world now, one so saturated by conflicting sources that many have lost trust in qualified voices such as doctors with decades of experience. If I save even one additional life by advocating for early detection with mammography—*not thermography*—all the roasting will have been worth it. And, of course, if thermography (or any other technology) undergoes future validation studies, with the same rigor that mammography and sonography have been studied, and demonstrates added value, we will incorporate it into our practices.

IF YOU FEEL SOMETHING, SAY SOMETHING!

KATHERINE ESKOVITZ, lawyer, author, and mom of three: I felt confident about my health after completing all my checkups in June:

- Blood work and breast exam with my internist—check
- Galleri early cancer detection blood test—check
- Annual mammogram and breast exam—check

Everything came back normal. I felt great. Three days later, I was lying in bed next to my husband, reading with my book propped near my chest, when I felt something—a hard pea in my left breast. I've replayed that moment in my head many times, and I still don't remember exactly how or why my hand landed there. I had never done a breast self-exam. I have no family history of breast cancer. I assumed I was likely immune.

I guided my husband's finger to it. "I just had a normal mammogram, so I'm not worried about this. But feel it—do you feel the hard thing?" After a few tries, he found it. "I have one of those in my arm. It's probably just a cyst or something like that," he said. I wasn't concerned, but I made a mental note to schedule an ultrasound—just to be safe.

Two days later, while running through a to-do list, I called the imaging center. The receptionist told me I'd need a referral since I just had a normal mammogram that week. I almost let it go right then—it must not be necessary to go in if they wouldn't see me automatically—but it struck me as odd that they wouldn't check a lump I could feel. Something in me insisted I follow through. I called my internist, and she sent the referral.

A week later, I went in for the 8 a.m. ultrasound. I hadn't thought about it all week and hadn't even mentioned it to my husband. I assumed it was routine.

The radiologist who had read my mammogram greeted me with an easy, untroubled manner that suggested he thought my visit was unnecessary. I suspected as much myself. But as he moved the wand over the spot I indicated, his expression shifted. He grew quiet, eyes locked on the screen.

"You seem concerned?"

"I *am* concerned," he told me.

He recommended a biopsy on the spot. I didn't know exactly what it involved, but his seriousness unsettled me, and I agreed.

I asked, "So what are we talking about?"

He told me that initially he was expecting it to be a cyst since he had just seen me, but it's good that I came in because it was a small tumor, either malignant or benign.

My mind felt like it was racing and frozen at the same time. I was trying to grasp how serious this was. I imagined my husband asking me a bunch of questions about the visit that I wouldn't know the answer to, so I forced myself to follow up with, "Is this a fifty-fifty situation?" I asked, trying to sound light, waiting for him to reassure me that most of the time it's benign.

But that's not what the doctor said. Instead, he said, "Based on the look of it and my experience, I think there's a 70 percent chance it's cancer."

I sat there on the table with my breasts exposed, tears prickling. I was not a person who got cancer. This could not be. Forty-eight hours later, it was confirmed: I had breast cancer.

I felt like a breast cancer impostor—diagnosed with something that didn't match how I felt. As I started to tell the people closest to me over the next few days, the reactions seemed so big. But I felt healthy. The flowers, the cards, the emotional reactions, it seemed like it was meant for someone really sick. And I didn't feel really sick. I kept telling more people of my diagnosis; saying it out loud helped me process it. Each time I said it, it felt more real.

My advice is that if you want to avoid the cancer club, or shorten your stay, listen to that quiet voice in your head, that whisper, when it tells you something is not quite right. After all of my annual exams were normal, I am still surprised that I pushed for an ultrasound. A friend later texted: "That's called bravery, Kath. I think a lot of people would've chosen to bury their head under the sand and accept positive news of the mammogram instead of getting it evaluated. It's a brave thing to face potentially bad news and you did it head on."

I don't feel brave, but I do feel grateful I listened to that voice.

DR. PORT: Neither the American Cancer Society (ACS) nor the United States Prevention Service Task Force (USPSTF) recommends breast self-exams as a form of screening. In fact, the USPSTF actually gives breast self-exam a D rating (on a scale of A to D), recommending against the practice and saying the harms outweigh the benefits. Try telling that to Katherine and so many other women who have found their own cancers in between yearly imaging studies, also known as "interval cancers," or to the 10 to 20 percent of women whose cancers are "mammographically occult," and don't show up on mammograms at all.

The USPSTF cites two main reasons for their recommendations against breast self-exams. The first is the claim that breast self-exam has never been shown to reduce mortality from breast cancer. I don't buy that as a good reason, since we know that early detection saves lives. Furthermore, given the USPSTF current recommendations of only every *other* year mammography, which I also don't agree with, why wouldn't it be safer to check ourselves in between? Waiting longer between screenings only increases the risk of finding a more advanced cancer, which could be picked up by a self-exam in between mammograms.

Their second reason for recommending against breast self-examination is related to the potential for increased anxiety and potential for biopsies. Of course, women are nervous about getting breast cancer. It is *the* most common cancer that women get. Telling women to calm down and not examine themselves is patronizing and paternalistic. The irony is that both organizations recommend "breast self-awareness" and "being familiar" with what is normal for you, as well as seeing a doctor if you notice any changes or feel anything concerning. It is counterintuitive to expect women to be "breast self-aware" and recognize changes while telling them not to perform self-exams.

When you compare the self-exam recommendations for breast cancer with skin cancer, the disconnect only gets clearer. I have a family history of melanoma. I do worry about getting it myself, given my family history, compounded by the fact that I grew up in Southern California, where we were out playing in the sun almost every day, lathered in baby oil or Coppertone SPF 0. My dad was one of those guys on a lawn chair with a reflector aimed at his face and bald head, trying to "get some color." To make matters worse, when we visited our grandparents in Florida, we'd get countless bathing suit–line sunburns building sandcastles, catching crabs, and, as we got into our teenage years, "lying out" like our dad. Knowing my risk, I do get two yearly skin checks, and I see my doctor for new things I notice, which often do develop in between exams. Guess what? The ACS and USPSTF recommend I check myself for skin cancer too! And not just me, the poster child for melanoma risk, *everyone*. Skin self-exam is a practice that the ACS recommends, and USPSTF gives it a C rating (insufficient evidence for or against; not a great rating, but better than the D the breast self-exam gets). So, let me get this straight: It is recommended to examine our skin, where we can also "find things" and end up with biopsies, but not our breasts?

The main rationale these organizations use to recommend *against* self-exam can be summed up in two misleading words: *fibrocystic disease.* This is a term used by many doctors, although

I prefer *fibrocystic changes,* as I don't think this is something that should be categorized as a disease, for reasons I will outline below. In order to understand what doctors mean by fibrocystic disease, let's loop back to breast density. Fibrocystic tissue is essentially a variation on dense breasts, something that is normal for many women. Fibrocystic changes are also completely normal for young, premenopausal women. Fibrocystic tissue can wax and wane cyclically, usually related to the menstrual cycle, and can result in increased tenderness, lumpiness, and even increase in size or swelling of the breasts due to hormonal stimulation. Patches of breast tissue or the whole breast can become inflamed or sensitive. It can even feel like there is an actual, more prominent lump that can be breast tissue or a cyst (a fluid-filled sac). These changes usually peak right before a menstrual cycle is coming on, only to regress in the week after. To be clear, fibrocystic changes are not a disease and shouldn't be called that, even though the condition is often referred to as such. The word *disease* implies there is potential harm that can result from this condition, when this is the normal baseline for many young women, starting in their twenties and sometimes lasting until their fifties. These lumps and sensations that come and go are not cancer, cannot turn into cancer, and do not increase your risk of cancer. Doing a breast self-exam if you have fibrocystic changes can be stressful in the beginning, especially if you feel all kinds of lumps or bumps. The breasts can feel like a cobblestone street. But I recommend examining yourself every month so you become familiar with what is normal for *you.* Then, if a new lump or mass develops that *may be* cancer, you can identify it as new or different against your normal background.

On a similar note, breast pain, which can be associated with fibrocystic changes, is usually not connected to breast cancer at all. Breast pain can be part of fibrocystic changes or even related to dietary factors like caffeine or fatty foods. My patients who have significant breast pain related to fibrocystic changes often ask what they can do about it. Reducing caffeine intake can be helpful, but for many this would be a nonstarter. So, I reassure them that this cyclical breast pain is not related to cancer, in almost all cases, and should not be a significant source of concern.

All this is to say: check yourself. Once a month or so. In the shower. In the mirror. One method for doing this is to pretend that your breast is the face of a clock, with the top of the breast, directly below the collarbone, being twelve o'clock. Start closest to the nipple, in the twelve o'clock axis, and go around in circles gradually outward, using a circular motion with your fingertips, pressing in as you go. Don't forget to feel under your arms, as you make your way outward. The breast, as an organ, is teardrop shaped, with the point extending upward into the armpit, so checking the breast tissue in that area along with the lymph nodes that live under the arm, is important too. If you do feel something, it's not unreasonable to wait a month, a cycle—if you are still getting your period—to see if it goes away; nothing bad enough to affect an outcome will happen in that short an interim. After that, if you do feel something, and it doesn't go away, speak up. Advocate for yourself, just like Katherine did.

A NOTE ON THE CURES ACT AND WHAT IT MEANS ABOUT TEST RESULTS

Regardless of the reason that brought you here to read this book, I think it's worth it to say a few words about the 21st Century Cures Act, which a few years ago completely changed the way American patients receive information related to medical care. Whether you are waiting for a yearly mammogram report, the result of a biopsy, or the surgery report, the Cures Act will almost certainly affect your journey as a patient and as a person in some way. Perhaps it already has.

This law was enacted to prevent information blocking and requires complete and immediate transparency and access for patients to their medical information and all test results. Signed into place in 2016 by then-President Barack Obama, this law took full effect in October 2022 and mandates that all results are released directly to a patient the minute they become available, often even before their doctor has had the chance to see them, and irrespective of a doctor's availability to discuss or review. While no patient should be prevented from accessing their medical information or results in a timely fashion, in my world, this law has had a profound impact on the way patients receive results of a "bad" mammogram, a report from a medical procedure or test, or even a diagnosis of cancer. These results include the pathology reports from their cancer surgery, scans, or imaging, all of which can be really great—and a huge relief—or really scary.

Remember, results of tests are not always issued immediately after the test is completed and can "pop up" as an alert in your electronic medical record, or MyChart, at any time of the day or night—hours or even days after the test was performed. A new cancer diagnosis coming through a patient portal at 7 p.m. on a Friday evening is never good: the doctor's office is closed, there is no one to call for reassurance, no opportunity to schedule follow-up appointments or to get your doctor's read on the results until Monday morning, which can feel like an eternity. Patients relay such upsetting stories to me every day: situations where the diagnosis of a cancer was made worse by the way they received the news.

In the cancer world, we grapple every day with how to best

deal with the changes this law has brought. Patients *do* have the prerogative to view their results, but doctors also cannot be available for such discussions 24–7. Remember, the delivery of these results, whether they are lab tests that are often automated, imaging reports issued by a radiologist, or biopsy or surgery results issued by a pathologist, is often part of a process whereby individuals are signing many, many reports each day, as part of their own work cue. The doctors generating these reports may be doing their work remotely, from home, or from the hospital at any hour of the day or night, so the timing of when results will be available is not usually predictable. For example, basic lab screenings like blood tests may be available within hours, whereas reports on specimens taken at surgery may take days. These doctors are typically not communicating with your doctor regarding results, and also do not typically communicate directly with patients, with whom they have no relationship. They "release" their completed reports into a computer system, which is now directly available for your viewing, in real time.

While I am always available for my patients' medical emergencies and I wish I could be available at any hour of the day or night for test result interpretation and discussion too, it's simply not realistic. For example, on my surgery days, I am scrubbed in for hours at a time. Not surprisingly, when in surgery, I am totally focused on that patient. So, I often won't see other patients' results until the end of the day, potentially hours after they have already seen the results themselves. I have had patients say everything from "I'm just not going to look. I will wait for you to call and go over my report with me," to "I googled every single word in my pathology report before seeing you," only to find out that their interpretation of their results was only partially helpful or accurate. Everyone wants the relief that comes with good results sooner, but no one wants to receive bad results ever, much less on their own. It's tricky, and it's personal. My advice would be to make decisions up front about how you want to receive your information, and on what terms. Are you okay receiving results earlier, but potentially dealing with bad news on your own, or would you rather wait to have it all put into context, with a plan to follow? As you can perhaps tell, I favor the latter. It's certainly hard to defer checking results when you have been notified that they are available to you; we all want the relief of good news as soon as possible! But the decision to check results or wait to hear them from your doctor is a decision only you can make for yourself, and you will have to make this decision many times over the rest of your life about so many aspects of your health care—not just breast-related—with the Cures Act now firmly in place.

THE BREAST SCREENING ADVICE

PATIENTS' WISDOM AS YOU APPROACH
SCREENING DECISIONS

HANNAH STORM, sports anchor, producer, and director: Families and friends need to operate as a team. I always tell men, "Help your wife get her mammogram on the calendar. Drive her if you need to, take her out to dinner afterward. Watch the kids if you need to." As working women, as mothers, as women who are taking care of our parents, as women in general, we put everybody else before ourselves. That's why it's important to develop a network. Make sure you are doing everything you need to do to encourage the women in your life to get tested.

I have a friend of mine who didn't have a mammogram for so long and finally *my* husband got irritated and said, "I will take you and your group of friends all out to dinner if you get your mammogram." So, she got her mammogram and texted him a picture of all of us afterward saying, "Where are we going for dinner?"

GINA S., media professional: If you are not comfortable with your provider, it's okay to go to someone else. If you feel something is off, your gut reaction is probably right. Never feel dismissed by a doctor. That's my best advice. If you feel like a doctor is not responding to your emails or giving you the time of day, if you feel dismissed by a doctor, time to get out of there. You're supposed to be the most important person to them in the time they are taking care of you. And you should feel that way.

MY ADVICE

If you are not at increased risk for breast cancer, start annual mammograms at age forty. Add ultrasounds if your breasts are dense. If you are at increased risk for breast cancer, added MRI screening as well as starting mammograms earlier should be considered. More on how to figure out your risk level and when to start these added studies in the next section on increased risk. Talk to your doctor before pursuing alternative types of screening tests. Perform monthly breast self-exams and always, if you feel something, say something. Be mindful of the possibility of false positives associated with any imaging tests; follow-ups may be required but are not commonly a cause for alarm. And if you're not getting the information you need about any of these tests or the guidance you need regarding which tests to have, try another doctor. Think through your options for reviewing test results alone versus with your doctor. Know yourself and what's best for you.

Finally, a small piece of insider's advice from me on the timing of any screening test: if at all possible, consider what else of importance might be going on for you at that time. After years of hearing women tell me that their "bad mammogram" came right before some important event for them, I always remind my patients that any test they do has a small chance of showing not perfect results. So, while there is *never* a good time to receive a call-back, much less get a diagnosis of cancer, it can be particularly difficult to get concerning results while something else of extra importance is about to happen in your life. Do you have a major presentation to give at work that could result in a promotion? Perhaps you shouldn't do your mammogram the day before. If the results are concerning or even just need further follow up, this can be a major distraction and extra source of concern. The week after might be better, and most facilities are able to adjust appointments and reschedule within a reasonable time frame. Your son's bar mitzvah is two weeks away. Maybe wait till after for that breast MRI so that again, if results require further follow-up, you and your family can enjoy yourselves without having these questionable results hanging over your head. This is one way to set yourself up for success in screening as it relates to timing.

HIGH RISK

WHAT MAKES SOMEONE HIGH RISK?

11

GENETIC TESTING

KNOWLEDGE IS POWER

CHRIS EVERT, former professional tennis player: The only reason I'm alive is because of the genetic road map my sister left behind. She had ovarian cancer, an insidious cancer because you don't have any signs. The only sign she had was that she was starting to get out of breath, and she was more tired than usual, but there was no pain, no blood, no nothing. We were traveling in Singapore and I told her that when we got back, she had to go and see her doctor to get tests to figure out what was going on. She got exams and scans and the doctor said she had ovarian cancer. When she went in for surgery to get a hysterectomy, surgeons saw that her whole body was riddled with cancer. It was stage IV. She passed away from it in 2020.

My sister was tested for a BRCA mutation. Everyone has a BRCA gene, but not everyone has a mutation in the BRCA gene. She tested negative but she had a variant of uncertain significance (VUS), which meant there was a difference in her gene but there hadn't been enough testing on it to know if it was pathogenic or cancer causing. At the time, they told us that since she was negative, we didn't have to get tested, so no one else in my family got tested.

Two years later, I got a phone call from my sister's geneticists and they said, "The variant your sister had has now been reclassified as cancerous. She was in fact positive for the BRCA gene mutation. I would advise all of your family to go out and get tested." We all went out and got tested and while some members of my family were lucky, my test came back positive.

I went to my sister's ovarian cancer doctor, and he said, "You need to get your ovaries out right away." Since my sister's death, I was getting internal ultrasounds, blood work, X-rays, everything. Without knowing if I had anything or not, I had the ovaries and fallopian tubes out with no questions asked, just to be preventive. Turns out I had cancerous cells in my fallopian tubes and one tumor in my ovary. My doctor was shocked. Just like my sister, I hadn't felt anything. My numbers had been good and there was no need to think that I had cancer. Ten days later, I got more surgery to remove the rest. Then, they placed me on chemotherapy.

After the chemotherapy, my doctor told me I had to make a decision whether or not to have a mastectomy. I no longer wanted to have three, four checkups a year. Knowing my family history and that I had a 60 to 80 percent chance of getting breast cancer without a mastectomy, it was not a difficult decision, so I went to see Dr. Port.

Two years after that, out of the blue, my doctor suggested we get a CAT scan of my abdomen, even though it wasn't my time to get a CAT scan—I was getting them every six months— and they found another tumor related to the ovarian cancer. It was unbelievable. So, I went through my second bout.

Looking back, we might have slipped through the cracks for genetic testing. We are not Ashkenazi Jews and have no family history of breast cancer, so we were the last people expected to have the mutation. *But* my mother had a hysterectomy when she was fifty years old. She was still getting her period, and she hemorrhaged one day at a tennis tournament. They didn't even do genetic testing back then. Or, the gene could have come from my father, so it's important to test both sides of your family.

I just wanted to scream out on top of the mountain about genetic testing, family history, and early screening. If I hadn't detected cancer, if I hadn't gotten screened, it could have progressed quickly to stage III or stage IV. For some reason, reproductive cancers have had a taboo reputation, but I felt no qualms about getting the message out because it is a reality and it's nothing to be ashamed of. I read that only 10 percent of women and men who have the BRCA mutation know it. You have to be your own advocate. If you feel anything different, even if it's as small as you ran a mile two weeks ago and it was easy, and now you're out of breath, go get it checked out.

I felt it was a responsibility to push all of my family to get tested: my cousins, siblings, children. I took it seriously because it's a family thing. Even those who didn't have the BRCA gene mutation were grateful because many of them found other things. I also had my friends get genetic testing. I thought it used to be a pain in the butt to have to fill out the forms at the doctor's office, like, "Why do they have to know about my grandmother?" Now I know that genetic testing can help people catch their cancers at an early stage.

People will say, "You're so strong, you're a warrior."

Here's the thing, when you're faced with something that's life or death, you just do it. You go through chemo, you go through surgeries, and you just do it. There's no other option. When I meet people at tennis tournaments, they don't just say, "I admired you so much as a player," they say, "Thank you so much for being outspoken about cancer. I had breast cancer," or "I had ovarian cancer." If I had kept it to myself, knowing I wasn't helping other women go through this, I think it would make me feel worse. It's the silver lining: knowing that something good always comes out of something bad.

Despite everything, I feel like I had an easy time. It's about your mindset. If you want to live, and if that's what it takes to increase your chances of living, it should be a no-brainer.

DR. PORT: Let's talk about genetics. In 1994 and 1995, two genes were identified called BRCA1 and 2. Scientists studied families with strong family histories of breast cancer—especially early onset—and ovarian cancer. All people have BRCA genes, two copies, in fact, one from each parent. The risk of cancer starts when there is a mutation in one of these inherited copies. Whether the mutation comes from the mother's or father's genes, it puts the carrier at a higher risk for developing a variety of different cancers. Approximately one in four hundred people carry a BRCA mutation. However, among the Ashkenazi Jewish population (Jewish people of mostly European ancestry) the risk is one in forty. So, among this Jewish population, 2 percent will carry the mutation. The discovery of these two genes was a monumental leap forward in our ability to understand and identify those at highest risk.

Shortly after they were isolated and identified, we were able to start testing individuals for mutations in BRCA1 and 2. The test is easy, just either a blood or saliva sample, and results usually return within two weeks. A test showing a mutation in one of these genes would mean that a woman would have a 60 to 80 percent risk of developing breast cancer over her lifetime and a high likelihood of developing it at a young age. The average age of breast cancer onset with *no* genetic predisposition is approximately sixty; for BRCA mutation carriers, meaning people who have the genetic mutation, it is in the early forties. The risk of ovarian cancer is also significantly higher: between 20 to 40 percent, compared to a low risk of 1 percent in the general population. BRCA mutations are also associated with increased risk for other cancers: prostate cancer in men, melanoma, pancreatic cancer, and male breast cancer. These less common cancers develop with higher frequency than the general population, but at much lower risks than the extraordinarily high risks of breast and ovarian. For example, the risk of pancreatic cancer in the general population is 1 percent. With a BRCA1 mutation, it increases to 2 to 3 percent, and to 5 percent with a BRCA2 mutation. While these risks are still comparatively low compared to the substantively higher risks of breast and ovarian cancers, pancreatic cancer is a serious and quite life-threatening disease. Even a small increased risk is a major health concern.

BRCA1 and 2 are the same in some ways and different in others. Both are associated with high risks of breast cancer. Both are associated with a high risk of getting breast cancer at a young age. Both are associated with higher risk of ovarian cancer compared to the general population. Both are associated with a high risk of pancreatic cancer and prostate cancer in men. Here's how the two genes differ. Newer data show that the risk of ovarian cancer may be earlier onset (at or around age forty) with BRCA1, compared to BRCA2 (at or around age forty-five). Preventive surgery, removal of the ovaries, is often the best option for carriers of BRCA1 and 2 mutations, and while it may seem wild to make such a move based on a blood test or finding a genetic mutation, there are no great screening options for ovarian cancer and detecting it early is a challenge. As such, preventive surgery is the most assured way of addressing such a high-risk situation with few options for early detection. While we may strongly advocate for BRCA1 mutation carriers to remove their ovaries *by* age forty, BRCA2 mutation carriers may be able to hold off on surgery for a few more years, retaining and prolonging normal ovarian function for additional time, and providing a slightly longer window for childbearing.

The two mutations are also associated with different subtypes of breast cancer, which have different treatment pathways and levels of aggressiveness. About 60 to 70 percent of breast cancers are hormonally driven, meaning that they're fed and stimulated by estrogen and progesterone—and that hormone-blocking agents are very effective in reducing the risk of their recurrence. If you carry the BRCA2 mutation and you develop breast cancer, odds are, it'll be the same kind of cancer that most women get. However, women with BRCA1 often develop a form of breast cancer that is called triple negative. This type of breast cancer is more aggressive and more complex to treat, since many therapies do not work on it. Triple-negative breast cancer comprises only about 15 percent of all breast cancers diagnosed in the general population. However, if you carry the BRCA1 mutation and develop breast cancer, there is a *75 percent chance* that it will be this kind of breast cancer. BRCA2 mutation carriers *can* develop triple-negative breast cancer, but it's far less common and closer to the 15 percent rate we see in the general population of breast cancer cases. Many women with BRCA1 mutations do opt for preventive surgery, because they are aiming to specifically prevent the aggressive form of breast cancer that they are at higher risk of developing compared to BRCA2 carriers and definitely compared to the general population.

Finally, the third difference between BRCA1 and 2 has to do with the risk of male breast cancer. Chapter 34 is devoted entirely to this very rare entity. Male breast cancer comprises only 1 percent of all breast cancer cases, and less than 1 percent of all cancers in men. In the general population, the risk for men to develop breast cancer is far below 1 percent. But men with the BRCA2 mutation have a significantly higher risk of developing male breast cancer: approximately a 7 to 10 percent risk level. Men with BRCA1 have a risk of approximately 1 percent. That 1 percent risk seems quite low, but the risk is still much higher than for the general population.

Because breast cancer is so common and can develop at young ages with BRCA mutations, testing for these mutations provides important health information regarding specific risks for cancer. It helps me counsel an individual patient about their risks. We can also advise more clearly on escalating screening to optimize early detection. For this extremely high-risk group, recommended screening starts at age twenty-five with yearly MRIs, adding yearly mammograms starting at age thirty. Importantly, we typically don't recommend genetic testing much before age twenty-five because screening would not typically start before that age anyway. We continue this screening regimen each year with both of these tests. Most importantly, however, mammograms and MRIs don't prevent breast cancer. They are done with the goal of early detection for those who develop it. But even the earliest breast cancer detected may need aggressive treatment and has a risk of mortality. So for women who want to be more proactive about actual prevention, there is also the consideration of risk-reducing surgery—bilateral mastectomy (removing the breasts on both sides)—in this extremely high-risk group. Some patients are nervous about having genetic testing, because they fear doctors would push dramatic surgery and that they won't have any other options. But there are other options, and no one should be forced or even encouraged to do anything drastic. In reality, doctors help their patients make informed decisions by explaining their risks and options. Mammography and MRI are effective high-risk screening modalities for breast cancer in BRCA mutation carriers, and these yearly tests can continue ad infinitum, or until a patient is ready to undergo a bilateral mastectomy to reduce risk. Much of the world learned about this type of surgery when the actor Angelina Jolie wrote about her experience having this operation to prevent breast cancer after learning that she herself was BRCA positive.

If you think you could be a carrier of a BRCA mutation, there are specific red flags to look out for (described below), but the guidelines for genetic testing continue to evolve. The National Comprehensive Cancer Network (NCCN) is the major organization for developing evidence-based guidelines for many cancer screening and treatment practices, and the NCCN guidelines for genetic testing have been repeatedly revised and updated, casting a wider and wider net, as it should be. Chris Evert is the perfect example: wider testing might have detected her mutation earlier *before* her sister, and she, developed ovarian cancer, allowing for prevention. For example, up until 2023, NCCN guidelines recommended that all women diagnosed with breast cancer under forty-five years old should be tested for BRCA mutations. In 2023, the line in the sand was moved, and the age was raised to be more inclusive, recommending testing all newly diagnosed under the age of fifty. The American Society of Breast Surgeons, meanwhile, recommends that *all* women diagnosed with breast cancer be offered testing. Why not? Cost and access can be barriers. Not all insurances cover genetic testing in all situations, and testing used to cost thousands of dollars, making it unaffordable for most. But the out-of-pocket cost for genetic testing has come down dramatically, costing somewhere between $100 and $250, depending on the testing company. And it's worth it. Genetic testing can provide game-changing information for patients *and* their families. Family histories can be incredibly tricky and even unrevealing, especially if the lack of family history has to do with the lack of women relatives. More on this in the next chapter.

As Chris advocates, if more widespread genetic testing had been done and a BRCA mutation been detected *before* her sister's cancer diagnosis, and her own, preventive removal of the ovaries could have been performed, sparing her sister's life, and Chris's tough course of treatment.

Now we know about a slew of other genes associated with a higher risk of breast and other cancers, like CHEK2, ATM, PALB2, p53 and others. When we perform genetic testing, we tend to test for mutations in a panel of genes, not just in BRCA1 and 2. These genes also have increased risk of breast cancer but are associated with other cancers too. Each of these genes is associated with different recommendations for management depending on what cancer risks they carry.

Importantly, women who have breast cancer who do *not* carry any of these mutations are not at any substantive increased risk for other cancers just based on being diagnosed with breast cancer alone. Many of my patients diagnosed with breast cancer, for example, ask if ovarian removal is something they should also consider based on a perception of substantial increased risk for other cancers, such as ovarian. I reassure them that once they test gene negative, there is no reason to recommend this, and that their risk for ovarian and other cancers is similar to that of the general population. The increased risk of ovarian cancer associated with breast cancer is only true for those with a genetic mutation that puts a woman at increased risk for both.

It's quite exciting to know that it is currently possible to prevent passing down a genetic mutation to one's offspring with modern in vitro fertilization (IVF) techniques. Let's say I carry the BRCA mutation. If I desire to ensure that I do not pass on this gene to my children, I can undergo egg harvesting in a fertility clinic. In the lab, my eggs are then fertilized with my partner's sperm. These embryos are then tested for the mutation. We can choose to selectively implant and bring to life only those that do not carry the mutation, which should be roughly 50 percent of all the embryos we produce. Many genetically related diseases can be eliminated this way, and the pain and suffering that go with them, if a woman chooses to do so. It is an option with currently available technology, and many of my patients who are BRCA mutation carriers choose to do this.

In summary, genetic testing is a critical component of our evaluation for women *with* breast cancer. But it should also be critical for women (or men) without breast cancer! And it definitely should be considered if you meet any of the conditions below:

- You are diagnosed with breast cancer
- You have an Ashkenazi Jewish background
- You have a family history of any of the cancers you learned about in this chapter: breast, ovarian, pancreatic, male breast cancer, melanoma, or prostate cancer
- You are adopted or know very little about your biologic family history on one or both sides

Finding out if one is genetically predisposed is important and provides actionable information. Knowledge is power. Genetic testing for all is the future.

12

WHEN IT COMES TO GENETICS, DON'T FORGET YOUR FATHER'S SIDE

JILL MARTIN, *TODAY* show contributor, entrepreneur: I was always super vigilant about getting screenings because my mother had breast cancer when I was in college and my grandmother died of breast cancer. It was always something I thought about. We have the same hands, the same way about us. I thought, *When will be my time?* Every time I went for a scan, I was holding my breath, but I was always on it. My father had never been to a hospital until he just got his shoulder done a year ago. He's eighty-one. I had never known sickness from my father's family and didn't know breast cancer from my father's family. When my mother had DCIS (the earliest stage of breast cancer), she got the BRCA test; she was negative, so I didn't think I needed genetic testing because she was negative.

Then very randomly, one of my doctors asked me, "Did you ever get tested for BRCA?" "No, but my mother was negative," I answered. "There's no breast cancer on my father's side." We later found out that my aunt Zora (on my father's side) did have breast cancer. This doctor told me to just get tested, and she sent me a spit test, and I spit in it in between Zoom calls and sent it in. Three weeks later, on June 26, I get this call telling me I'm BRCA positive. The rest of that five-minute conversation was a blur. I do remember them saying, "We suggest you go in, we suggest you get an MRI, preventive surgery," the whole thing. That was the end of "BC," before cancer. I have calendarized my life before and after that phone call.

I learned I have a 60 to 80 percent chance of getting breast cancer. The first thing I spiraled into was, *My goodness, my mother's test wasn't right because it was done in the '90s. It must be positive and she doesn't know. Now I'm nervous that my mother is walking around with the BRCA gene.* Both my mother and father got tested. In the end, it turned out it was my father who was positive, which was shocking and heartbreaking. I felt devastated for him. He gave me so many beautiful qualities and such a beautiful life, but he also gave me this gene.

I had a normal mammogram a few months before, so I felt like I was okay. When I saw Dr. Port, she listed all of the things that you have to do when you're BRCA positive. Then she examined me, and she felt something, and her instinct was right. She said, "I want you to get an MRI immediately." And that would have happened before the surgery anyway, but it was July Fourth weekend, and so it was a messy time to do that. I remember I was in the car, and we were going to do it, and it was just a rush. I went and I got the MRI, and I walked out, and I wasn't supposed to see the screen, but of course with my personality, I walked over to it, and I said, "What is that?" I saw a huge mass. I'm not a doctor and at the time I didn't know what anything was, but I knew it was something. I went to get a biopsy, and it was cancer.

Doctors I've asked have told me that in another year it might have spread and been incurable, because it was on the move: in real time. When I woke up from my double mastectomy, Dr. Port told me that she found that it had already spread to one lymph node. Knowing the guidelines, I was certain this would lead to chemotherapy, which is the only thing that people really talk about because it's horrible. It's something I think about every single day and that I'm traumatized by. Chemotherapy was a lot, but I got through it. Then I got radiation. Then I took my ovaries and fallopian tubes out. Then I had another reconstruction. Now I'm on two "safety net" pills.

When people say, "You know, in a few years, it'll be behind you," I say it won't be because I am choosing for it to be my identity for a few reasons. I choose to shout about genetic testing from the rooftops. You should get it, of course if you have a family history, and remember to pay attention to your father's side, but even if you don't. If everyone got tested, then if they found out they had the gene, they could get screened to catch it early or have the preventive surgery. This is my advice, and the message that I want to amplify. I can't let one more person go through what I went through if there is a way to avoid it. If there is an option to do this game-changing preventive surgery, you can move on with your life. With cancer, you don't *move on* with your life. You move on with it in a different way, in so many beautiful and magical ways.

DR. PORT: In Jill's case, she had a family history of breast cancer, but the genetic component was surprisingly from the other side of her family, her father's. By contrast, my patient Sophie came in at thirty-two years old with a new diagnosis of breast cancer, and no family history at all.

She had noticed a lump in her breast approximately two weeks prior. She was about to take the bar exam, so she waited to see a doctor until it was out of the way. The day after the exam, she called her gynecologist, who ordered test after test, leading to the diagnosis. Now, instead of starting her law career, she would be starting chemotherapy and undergoing treatment for breast cancer.

"How is this happening?" she asked me.

"I'm not sure yet, Sophie," I said apologetically. "Some young women develop breast cancer, and we never find out why, but genetic factors and specific genes can cause early onset breast cancer. Let's talk about your family for a minute."

"I have no family history," she answered curtly. "My mother is alive and well. So are her sister and her brother. I have an older brother who is totally fine."

"Okay," I said. "Now let's talk about your father and his side of the family."

"What does that have to do with anything?"

"You inherit genes from both sides of your family, even genes for breast cancer."

"My father is alive and well," she said. "He's out in the waiting room, in fact."

I dug a little deeper.

"What about his siblings?"

"He has a brother who is fine as well."

"No sisters?"

"None."

"So, the truth is," I said, "we can't really *know* if there is any family history of breast or ovarian cancer on that side of the family if there aren't any women. Your father's side of the family could be harboring a gene that's stayed silent. BRCA mutations can hide in male lineages; some men get prostate cancer, it's a common enough cancer that it doesn't set off any genetic alarm bells. We may not know a family carries the gene until a daughter is born and develops breast cancer as a young woman. We need to get you tested."

Sure enough, Sophie tested positive for a BRCA1 mutation.

Even without stories about a mother, sister, or aunt who had breast cancer, my BRCA radar goes off when I hear the following:

"I don't know anything about that side of the family; we're estranged."

"My father has no sisters."

"I am adopted and know nothing about my biological family's history."

Let me give you a more personal example. In 2014, about fifteen years into my career as a breast surgeon and almost twenty years into my marriage, my mother-in-law was diagnosed with late-stage ovarian cancer at the age of seventy-five. I walked her into the OR for her surgery, which was to be performed by a close colleague who is an expert gynecologic oncologist. She had late-stage disease, as most ovarian cancer is diagnosed at more advanced stages, but thank God, she also had an aggressive and talented surgeon who was prepared to give her the best possible chance of a cure. When that doctor came out of surgery, he said they found disease everywhere, all throughout her abdomen. Still, he believed he had gotten it all, painstakingly dissecting out every stud of disease he could find on any organ surface. Like most women with ovarian cancer, my mother-in-law's disease was not caught early. It had grown and spread surreptitiously in her belly for a long time before she developed the classic symptoms of bloating and some abdominal pain.

She recovered at our apartment, so we spent a lot of time together over the following weeks. At some point, it dawned on me that, like the rest of our family, she was an Ashkenazi Jew with ovarian cancer and that this was a potential red flag for the BRCA gene mutation. My concern was somewhat mollified by the fact that she had made it to seventy-five without breast cancer, which is the more common disease for BRCA mutation carriers to get. But I thought about her family history: two brothers, no sisters. One brother had died at nineteen in a tragic accident, and the other did have some form of aggressive cancer. No one knew what kind, since it was so advanced at diagnosis. Maybe colon? Maybe stomach? *What if it was pancreatic?* When she was back home in Florida, receiving her chemotherapy, my husband and I called to check in on her. At the end of our conversation, I casually slipped in, "You know, Irene, I think you need to have genetic testing. If you were to test positive, it could have huge implications for your kids and ours."

"Okay," she said, seemingly ambivalently, "I'll talk to my oncologist."

A week later, we checked in again. At the end of our call, she volunteered, "So I spoke with my oncologist here in Orlando, and he said I didn't need to be tested, since I have never had breast cancer and have no family history."

I explained the potential for gene transmission from the paternal side in a family with no females. When I stressed that it could have huge implications for my husband and our kids, especially our daughter, she acquiesced and promised to do it. Three weeks later, the results were in: she had a BRCA2 mutation. That was the cause of her ovarian cancer. Now, my husband needed to be tested. As we waited an agonizing two weeks for the results, I went to sleep every night filled with dread and intrusive thoughts about what a positive test result would mean for us as a family, and specifically for our daughter, who, at that point, was fifteen years old.

"Listen," my husband said. "You do this every day. You empower people to step up and get tested, get screened and get preventive surgery. And you have been doing this for fifteen years now. If it's now us, we are going to walk the walk."

I couldn't believe the irony that after so many years of teaching others about genetic predisposition and how to manage it, we were now, suddenly, unexpectedly, potentially dealing with it ourselves. He was right.

"Every family has something," he continued. "And all parents pass down genes, good and bad, to their children. At least this is something we can test for! And, if we have it, it's something we could do something about."

It wasn't, "Why us?" but rather "Why *not* us?" Of course, it is easy to tell this story now. His test results came back negative, which means my mother-in-law's genetic mutation had not been passed down to my husband. It also meant our kids were off the hook; you can't pass down what you don't have, and genes can't skip a generation. Those few weeks in limbo gave me a small taste of what my patients go through: the waiting and anxiety over test results and what they might mean for us.

All of these stories of case after case and family after family, drive home Jill's point—and mine—about paying attention to both sides when assessing your family history. If there are few women on one side of the family, know that *not* having a family history is relatively meaningless, and consider getting tested.

Even if you do have a family history of breast cancer and that side has tested negative, in a strange twist, the other side could be positive. Here's my advice. Do a deep dive into your family medical history:

- Did anyone have cancer? If so, what kind?
- Was anyone exceptionally young at diagnosis?
- Does one side of the family (or both) have very few relatives, especially female relatives, so breast and other cancer risk can't really be assessed?
- Are you of Ashkenazi Jewish descent on either side?
- Is there an estrangement in the family, where one side doesn't know whether the other side has developed any cancer?
- Are you or either of your parents adopted with no knowledge of biologic family medical history?

If any of these situations ring true, get genetic testing. Jill chose to make her story public, and together we will amplify what she learned: that getting genetic testing and finding out that you are BRCA positive can be quite heavy, but it doesn't force you to do anything. You can be screened. You can be examined. You can also choose to do preventive surgery, which virtually eliminates the chance of getting breast or ovarian cancer. Ultimately, the choice is yours. Regardless, genetic testing empowers us to be able to make those choices. If even one unsuspecting person finds out they are gene positive and the result is the prevention or earlier detection of breast cancer, Jill's shouting from the rooftops for genetic testing will have been well worth it.

13

OTHER RISK FACTORS FOR BREAST CANCER

FAMILY HISTORY, ATYPIA, LCIS, AND BREAST DENSITY

JENNA P., mother and dermatologist: My sister had breast cancer at age fifty-five. I'm her age now and know that I am at risk. She had genetic testing when she was diagnosed, and so did I, but neither of us has any genetic predisposition. To cover all of my bases, I get screened every year with both mammograms and MRI. When you have a family history, you know you're at a higher risk of developing cancer. But when you don't know what the exact risk is, it's hard to know what to do. I have been told my risk is anywhere from 25 percent to 40 percent. That's a big range! Is screening enough? Should I be thinking about removing my breasts preventively? I might do it, if my risk was closer to that 40 percent number, but it seems a little extreme at 25 percent. For now, I am just continuing my screening and hoping that I don't get it.

MARLENE C., retired, grandmother: I had calcifications on my most recent mammogram. As it was explained to me, they are little white dots, like grains of salt of different shapes and sizes. I was told I needed a needle biopsy because they could be cancer. I had the biopsy. It was a bit painful, and uncomfortable, but of course I wanted to know either way. That's where it got complicated. I thought the results would be clear-cut and come back as either normal or cancer, but they came back showing "atypical cells." These cells, I was told, are funky looking, and while they're not cancer themselves, can live in the neighborhood of cancer. That meant there was a chance that I had cancer there *now*, but the biopsy just didn't fully sample the area.

My gynecologist, who ordered my mammogram, sent me to see Dr. Port. She and I discussed the options: either leave them alone and check again in six months or take the area out with a small surgery. She told me that the chances of finding cancer through this small surgery would be about 15 to 20 percent. If the calcifications changed or increased, I would definitely need them out; if they didn't, I could avoid the surgery. Ultimately, I wanted to know for sure, so I chose the surgery. The results showed more atypical cells, but thankfully, no cancer. Now at least I know.

Doing the surgery alleviated my anxiety. Now I know moving forward that the discovery of these atypical cells, in and of itself, means that I am at higher risk of developing breast cancer in the future, even though this wasn't it. My risk is now elevated to about 20 percent. Before we knew about the atypical cells, it was average, or about 10 to 12 percent. That change was important to know. And while

I am balancing the anxiety of knowing I am at higher risk for cancer, I also have the comfort of knowing and being able to pursue more aggressive screening going forward. Finding out that you are at higher risk for breast cancer is concerning, but there are options. Just like any other health situation, for me, it was better to know; both that I didn't have cancer and that I am at higher risk.

DR. PORT: Let's talk about the factors that might put someone at substantively higher risk, higher than the average woman's baseline risk of 10 to 12 percent, for getting breast cancer. In the following two chapters, we will go over how we put those risk factors together to wrap our heads around calculating that level of risk—and what we can do about it.

FAMILY HISTORY

While most women who get breast cancer have *no* family history of the disease (that's right, you read correctly), having a family history, as we've discussed, does put you at a higher risk. There are four factors of family history that go into our assessment of risk:

1. The closeness of the relationship with whoever in the family had breast cancer. A first-degree relative like a mother, sister, or daughter confers a higher risk to you than a second-degree relative, like an aunt, grandparent, or half-sibling.
2. The age at which that relative developed breast cancer, since a younger age is associated with higher risk.
3. If that person had breast cancer in just one breast or both.

4. Whether any male relative has gotten breast cancer. Yes, men can get breast cancer, but it's rare. More on male breast cancer in Chapter 34.

When we ask patients questions about family history, we use these factors to dial up or dial down our calculation of risk.

Having more than one relative who had breast cancer certainly increases risk. I have treated families where three successive sisters have all gotten breast cancer, with no identifiable genetic predisposition. *What's going on here?* we ask ourselves. I wish I knew. At some point in the future, we may know precisely why a family like this has multiple relatives with breast cancer. For now, they just have a pronounced family history. On the other hand, it can be hard to assess for family history when little is known about one or both sides of the family, as was discussed in the genetic section. What if you are adopted? Needless to say, you may know very little about your biological family history. Similarly, if one or both parents are only children, a small family may not reveal a lot of history, so it can be tricky.

BREAST BIOPSIES

Beyond family history, what other factors might increase the risk of breast cancer? Let's loop back to the beginning where I explained that if a person has an abnormal or new concerning finding on an imaging test, be it a mammogram, a sonogram, or an MRI, she usually needs a biopsy to find out what's going on, and a needle biopsy is the standard of care. Usually, based on these specimens of tissue, we can tell if the area is benign (normal) or malignant (cancer). Here's where it gets a little complicated, as I will explain below.

ATYPICAL HYPERPLASIA OR ATYPIA

The biopsy results don't always conclusively show normal cells or cancer. There is another possibility, a third "in between" category, where the cells show changes that are not *cancer* but are not entirely *normal*, either. When the findings fall into this "in-between" category it can mean different things. Maybe it means that even though this biopsy did not show cancer, the person now has an increased risk of *future* breast cancer. The results can also show that the cells are suspicious enough that they should be removed. One example of an in-between finding is atypia or atypical hyperplasia, which means that cells are, as the name says, *not typical*. Atypical cells look different from normal cells under a microscope. They show changes that *lean* toward cancerous type changes but have not reached that threshold. As such, we call this entity atypical hyperplasia or atypia.

Atypical hyperplasia can either be atypical *ductal* hyperplasia (ADH), or atypical *lobular* hyperplasia (ALH). As with Marlene, surgeons often suggest removing an area that has shown atypical cells, since there is a chance there will be actual cancer in this area, and this is true more for ADH than ALH. The discovery of cancer upon removal, which we call the upgrade rate, is usually approximately 10 to 20 percent with ADH, and less than 5 percent with ALH, which is why we are often more proactive about removal with ADH.

There are also times where we perform a biopsy that shows a papilloma, as discussed in the normal findings and biopsy discussion in Chapter 8. A papilloma in and of itself is almost always normal but, for a papilloma *with* atypical cells in it as reported on the biopsy results, we often recommend removal, as the upgrade rate here is also approximately 20 percent. It is a very individualized decision: to remove or not to remove. But also, just *having* atypical cells means that you are at higher risk for developing breast cancer in the future. The risk of getting breast cancer related to atypical hyperplasia is about 20 percent, and therefore increased above the risk of the general population. Importantly, even if patients get the small part of their breast with the atypia removed, that doesn't make their chances of getting cancer in the future lower. It just covers the base of finding out if a cancer might already exist in that area, seeing if there is any cancer in that spot *now* that the needle biopsy might have missed.

LOBULAR CARCINOMA IN SITU

Like atypia, lobular carcinoma in situ (LCIS) is also associated with increased risk. Yes, the name has *carcinoma* in it, but it's not cancer. As you can imagine, this misnomer often creates confusion and fear among patients who see these words on their reports. While LCIS is not cancer itself, like atypia, it also confers an increased risk of breast cancer, at 20 to 25 percent. We don't recommend removal of LCIS (again, achieved through a small surgery to that area) as frequently as we do for atypical ductal hyperplasia because the upgrade rate is much lower—less than 5 percent in most scenarios. Some women will say, "I don't care if the risk of finding cancer is 1 percent. I need to know, and I want it out." We can do that. The risk of the surgery is small, and some people are just not comfortable with any uncertainty and can't tolerate waiting six months for the next round of pictures to prove that all is stable. Still, no doctor wants to do, and no patient wants to have, unnecessary surgery, so if the risk of missing cancer is very low, it is very reasonable not to remove and follow up closely. Surgeons like me who specialize in breast surgery incorporate a number of different factors into advising our patients regarding removal versus watching and waiting.

- How big or small is the area of concern to begin with?
- If it was big, did the needle biopsy adequately sample what came out?
- Does the area look more concerning, or less concerning, than the results seem to show?
- Was the area "severely" atypical? Sometimes a pathology report will say that.

Our patients' preferences are always factored in, and we make this decision together.

CHEST WALL RADIATION

Receiving chest wall radiation, usually for the treatment of another type of cancer, can dramatically increase one's risk for future breast cancer. This is especially true when treatment is given at a young age, when breast tissue is still developing. The most common cancer requiring this treatment is a type of lymphoma, called Hodgkin's lymphoma or Hodgkin's disease, but there are other, thankfully rare, conditions where chest wall radiation might be given. The cure rate for Hodgkin's lymphoma is very high, but it may involve a type of radiation treatment called mantle radiation, which is given to eradicate lymph node disease within the chest cavity. While this therapy is quite effective and curative for Hodgkin's, receiving chest wall radiation, particularly at a young age when many with Hodgkin's Disease develop it, does put a young woman at significantly increased risk for future breast cancer. For this group of young women, high-risk screening is recommended, beginning approximately eight to ten years after treatment, but not usually before age thirty.

BREAST DENSITY

Breast density is complicated. Dense breasts are normal for a significant proportion of the population: approximately 40 to 50 percent of women, depending on age group. The only way to determine breast density is by having a mammogram. Mammograms define breast density, not how they feel or look to you, or even a doctor. Breasts that feel "lumpy," "cystic," or even hard may not necessarily be categorized as dense breasts. So how can increased density, which is normal for almost half of the population, also constitute increased risk? Dense breast tissue means there is a greater overall volume of the type of cells—ductal and lobular cells—that can turn into cancer. There is a lot of controversy over the degree to which breast density, in and of itself, increases risk of breast cancer. Suffice it to say that women with extremely dense breasts, approximately 10 percent overall, have some elevated risk above those with fatty, non-dense breasts. This elevated risk of developing breast cancer is different from the fact that separately, having dense breasts can also make it more difficult to detect breast cancer, as discussed earlier in Section 2. Having dense breasts is a bit of a double whammy: increased risk of breast cancer, *and* cancer being harder to see if you do get it.

As discussed in Section 2, if you have dense breasts, you should consider adding imaging beyond mammography. There is no one-size-fits-all solution, but the addition of a sonogram should be discussed, and an MRI can be considered, especially if you have other concomitant risk factors, like a family history. To be clear, having dense breasts *alone* does not elevate your risk to the 20 to 25 percent level that atypia and LCIS do, or that having a family history can. On the other hand, having extremely dense breasts, in combination with one of these aforementioned risk factors, can further add to your elevated risk.

HORMONAL FACTORS

To revisit hormonal factors influencing risk for breast cancer, there are a few: some we can control, and some we can't. Generally, more years of hormonal exposure translates to a higher risk. Getting your period before age twelve increases risk. As discussed previously, the average age of menarche has slowly but steadily decreased over the last century. So does late onset of menopause, when hormone levels drop. We cannot control the age at which we develop our periods or the onset of menopause. These are just factors to know. Thankfully, any added or decreased risk of breast cancer related to these factors is marginal. Getting your period at age ten, in and of itself, would not define you as "high risk" for breast cancer.

There are other hormonal factors within our control, to some degree, that also dial up or dial down breast cancer risk marginally. Given their minimal effect, it's therefore, in my opinion, not appropriate to make dramatic life decisions with them in mind. For example, not bearing children (also called nulliparity) is associated with higher risk of breast cancer than having children late in life, and having children late in life is associated with higher risk than having children at a younger age. Crucially, the risks associated with whether or not you have children and, if so, when, only adjust your risk marginally, by a few percentage points at most. It's estimated that for every five years' delay in childbearing it may increase your risk of breast cancer by 10 percent over baseline risk—the national baseline risk being 10 to 12 percent. Let's be clear: an increase of risk by 10 percent does not mean risk goes from 10 percent to 20 percent, which would be a huge leap. Ten percent *of baseline risk* (10 percent) is only 1 percent. So that's the added risk: 1 percent, not 10 percent. It's so important to understand this concept when determining one's risk and the factors that affect risk.

Interestingly, there is some data to suggest that among Black women, childbearing may have the opposite effect: increasing breast cancer risk with *younger* childbearing. This hypothesis might in part explain the concerning increased incidence of breast cancer in Black women in younger age groups—more on this later. Finally, breastfeeding, especially at a young age, has been shown to be protective against breast cancer, most likely by interrupting hormone cycles, given that most women who breastfeed delay the return of their menstrual cycle, which stopped initially with pregnancy. As mentioned in prior chapters, taking hormone-based birth control or hormone replacement therapy can also increase risk.

Overall, hormonal factors—childbearing, when you start and stop menstruating, taking or not taking hormones or birth control pills, breastfeeding—do affect breast cancer risk, but only minimally. By themselves, they would not place someone at high risk for breast cancer even when combined. My advice is to make the best choices for yourself with the totality of your overall health and life choices as the top priorities. But it is worth keeping an eye toward breast cancer risk as well.

BREAST CANCER RISK ASSESSMENT TOOLS

SCREENING AND PREVENTION STRATEGIES FOR THOSE WHO ARE AT INCREASED RISK

CHRISTINE E., supermarket manager: I'm fifty-two years old. At a recent appointment with my primary care provider, we started talking about breast cancer risk. I get a mammogram and a sonogram every year because I know my breasts are dense. Plus, my sister had breast cancer, so I already knew my chance of getting it was elevated; I just didn't know to what extent. My sister and I both did genetic testing, which was negative, but a lot of women get breast cancer who don't have the gene; my sister didn't. Our mother never had breast cancer, but she has other stuff like heart issues, probably because she smoked for years. As we talked, my doctor told me that there are ways of determining my risk: risk models. We went through one of them online together and found out that my risk of getting breast cancer is actually close to 25 percent, significantly higher than the average population. It's not that I didn't know my risk was higher, but it's one thing to know and another to put a number on it. Putting the number on it helped me have a more focused conversation with my doctor about what my options are such as getting more screening with MRI, which, for some women at higher risk, is added value in early detection.

My advice is that even if you are under age forty and not yet ready to start screening, know your risk. Because you might want to start screening earlier. Or do extra tests. Or talk about what you can do for prevention.

DR. PORT: Christine, like many women with a family history of breast cancer, knew she was at increased risk because of her sister's diagnosis. They both underwent genetic testing and were negative, so the cancer that her sister had was not associated with an identifiable gene. Importantly, that is true for most women with a family history. Only 10 percent of patients with a family history of breast cancer test positive for a known genetic mutation that puts them at the highest quantifiable risk of getting breast cancer, 60 to 80 percent, as you learned in Chapter 12. The remaining 90 percent of patients with a family history have no identifiable or specific cause of breast cancer that we can pinpoint.

This means Christine's risk is certainly higher than that of the general population with no family history (10 to 12 percent) but it's not as high as that of a mutation carrier (60 to 80 percent). So where does a person with a family history or a biopsy showing atypia or LCIS or dense breasts or all *three* of these risk factors fall? It's certainly somewhere in between 10 and 80 percent, but that's a large range! Is it 20 percent? Forty percent? Is there even a way to estimate risk for an individual with one or more risk factors for breast cancer? The answer is yes. We have various statistical risk models called breast cancer risk assessment tools that we can use to provide information on approximate risk. These models use statistical analysis to estimate risk based on an individual's profile factors that are plugged in. These models are all available online, and can easily be accessed and you can run them yourself or, if needed, with the assistance of a doctor.

The Gail Model and the Tyrer-Cuzick or IBIS model are the most well-known assessment tools, but there are others, as well. Each one factors in different information to calculate risk. Importantly, the different models incorporate and prioritize different risk factors. For example, some risk models factor in breast density; others don't. Some models only ask about first-degree relatives with breast cancer; others take a deep dive into family history, asking about first-degree relatives, second-degree relatives, *and* all of their ages at diagnosis. Given these different emphases from the different risk models, it's not surprising that the estimated risk results are not always the same. If you take Christine's case and use the Gail Model to calculate her risk, the results estimate that her lifetime risk of getting breast cancer is approximately 16 percent. We'd consider this a slightly elevated but intermediate level risk. Conversely, if you use the Tyrer-Cuzick model, which factors in her extremely dense breasts *and* the young age at which her sister developed her breast cancer, Christine's risk is estimated to be 23 percent. This discrepancy of seven percentage points may not seem so dramatic or significant, but it is. Twenty percent risk is a significant threshold—a line we draw in the sand, if you will—that defines the difference between elevated risk and high risk. And this is a level of risk that is considered actionable, both for added screening and prevention. Since the models provide different answers, which one should you believe? Unfortunately, it's not always clear. Fortunately, the range of risk levels between the different models is usually not that wide. And risk modeling is most useful in helping initiate a conversation with your doctor or a breast specialist about elevated risk if you are found to have it. The important thing is the overall idea that these models can help quantify risk, put a number on it, as Christine says, which can be helpful as a starting point for a discussion with your doctor about *individualizing* your screening, prevention, and breast health.

SCREENING FOR HIGH-RISK SITUATIONS AND WHEN TO ADD ADDITIONAL TESTS

Now that we know how to quantify risk, how do we act on this information? One thing we can do is adjust screening recommendations. To see what that might look like, let's revisit the different screening modalities but now as they apply to high-risk populations.

1. Mammograms are the basic standard of breast cancer screening. All women who have no increased risk should start them every year beginning at age forty. If a woman is at increased risk for breast cancer based on her family history, we consider starting mammograms earlier, often ten years before the age her youngest relative was when she was diagnosed. For example, if my mother was diagnosed with breast cancer at age forty-eight, I should consider starting to get yearly mammograms at age thirty-eight. That said, notably, mammograms are not typically recommended before the age of twenty-five, even in the highest risk populations.

2. Sonograms are now frequently recommended as an additional screening measure for women with dense breasts, regardless of their risk, so it's a no-brainer to add sonograms for women at increased risk, almost regardless of breast density. As discussed earlier, the added value or likelihood of picking up a cancer with an ultrasound in a woman with non-dense breasts is minimal, but with no added radiation exposure, there isn't much of a downside either.

3. For screening in high-risk populations, MRI is the differentiator. With its high sensitivity, MRI screenings are considered standard of care for the highest risk patients, genetic mutation carriers and those who received prior chest wall radiation, and should also be considered for other high-risk patients at the 20 percent risk level or above. MRI should be discussed with a patient's primary care provider when your risk is above 20 percent. Importantly, this is the risk level above which insurance companies will usually agree to reimburse for the test, which is important to factor into decision-making.

As discussed in previous chapters, for BRCA mutation carriers, the high-risk screening guidelines are clear: start MRIs at age twenty-five, and add mammograms at age thirty, doing each of these once a year. But, if one is not a mutation carrier, imagine how confusing it might be for a patient (and her doctors) to decide whether or not to add MRI when different risk models calculate different levels of risk. This is where calling in a specialist, someone like me, can be helpful. A breast cancer specialist, usually a surgeon, can provide more detailed, nuanced information about the risks and benefits of having an MRI and, of course, advise if the results are not completely normal. A breast specialist might look at individual risk factors in detail, make sure the patient has undergone genetic testing if appropriate, and then decide which risk model might be most fitting. If a patient has a family history that includes someone being diagnosed young, the Gail Model might *underestimate* risk, given that it doesn't ask for age at which the family member was diagnosed; having a family member diagnosed at seventy does confer a different risk than a relative diagnosed at age forty, and the Gail Model doesn't differentiate, but the Tyrer-Cuzick does.

As helpful as our risk models are, they do have limitations. They may underestimate risk for Black women. They aren't appropriate for mutation carriers, who are already at the highest risk level, and will not further quantify or adjust risk in a patient that we already know has a risk of 60 to 80 percent. Most risk models are also not relevant for women who have a history of LCIS. As discussed, LCIS puts a woman at about 25 percent risk level, so this factor *alone* means that she meets criteria for MRI screening. The risk models also can't determine risk of a *future* second cancer in someone who has already been diagnosed. If you access one of these risk models online, one of the first questions asked involves ascertaining a history of LCIS or an actual prior cancer, and stops you at "yes," meaning the risk model is not valid in these circumstances.

Like Christine, most women already know that they are at increased risk before ever running a risk model on themselves or asking their doctor to do it with them. There are two factors and only two that can get you to high-risk status: a family history or a high-risk pathology, meaning LCIS or atypical hyperplasia. There should be no *major* surprises. If you don't have a family history of breast cancer or one of these pathology diagnoses, there is no way a risk model will find you to be above 20 percent risk. Even if you have *all* the other minor risk factors combined—dense breasts, started your period early, and others—no combination can get you to high-risk status without a family history or biopsies showing high-risk findings. It should be comforting to know that as long as you don't have a genetic mutation or any known risk factors, a visit to your doctor's office will not lead to learning out of left field that you are at high risk for breast cancer.

On the horizon are *new* AI-based risk models. These novel, recently FDA-approved tools are using information based *only* on a patient's recent mammogram to predict risk of breast cancer in the next five years, without knowing anything about a woman's family history, previous biopsies, or any other fertility or hormonal risk factors. There are many other AI-based tools being developed around predicting risk that go beyond traditional risk factors. When it comes to imaging and prediction of risk for breast cancer, AI is the future, and the future is now.

So, finding out that you are, in fact, at high risk for developing breast cancer should open up the conversation for screening: when to start, which tests to do, and how often. Remember: screening is not prevention. Mammograms, sonograms, and MRI give us the best chance of picking up breast cancer early and often prevent late-stage breast cancer diagnosis through early detection, but they don't prevent the development of breast cancer itself. Is breast cancer preventable? Below are two strategies for breast cancer prevention, other than the lifestyle and risk modifiers discussed in the first chapters.

CHEMOPREVENTION

Some of the antihormonal medications that women take for breast cancer treatment reduce the risk of new cancers developing. About twenty-five years ago, when clinical trials clearly demonstrated that they have preventive effects in women at high risk, we started offering them to these women to reduce their risk. The most common medication is called tamoxifen, which we recommended for a course of five years, and which continues to have protective effects for another five years after taking it. Tamoxifen reduces breast cancer risk in high-risk women by approximately 50 percent. This means that if your risk is 20 percent, taking tamoxifen would cut your risk down to 10 percent, which is about the level of risk in the general population. In some patients, specifically those at high risk based on their diagnosis of atypical hyperplasia, risk reduction can be as high as 80 percent, meaning that it could reduce risk to a level *lower* than that of the average risk woman.

These results beg the question: Why wouldn't *all* high-risk patients take tamoxifen? We thought women might demand tamoxifen for everyone when the findings were published in 1999. That's not what happened. In fact, many women who were offered tamoxifen to reduce their high-risk status were reluctant to take it, primarily due to the side effects, which are all well described: potential for menopausal symptoms, blood clots, and uterine cancer, to name a few. While menopausal symptoms, such as hot flashes, are undesirable side effects for the women who experience them, many are, of course, more concerned about the more serious ones even though they are quite rare. Yes, blood clot risk is increased, but only by less than 1 percent— similar to the risk of taking birth control pills, which many women take without thinking twice. Similarly, while the risk for uterine cancer is also elevated, it's also by less than 1 percent, and it's even more rarely seen in women under sixty.

Initially, tamoxifen, and its cousin, raloxifene, were both underutilized; only an estimated 20 to 30 percent of high-risk women took them for preventive purposes. Even among women who did agree to try one of the drugs, many stopped taking them due to side effects. But here's the new development. In 2023, an Italian group published the results from a large clinical trial showing that a lower dose of tamoxifen (5 milligrams versus the standard 20 milligrams per day) for only three years instead of five was also extremely effective at preventing breast cancer in high-risk women. The lower dose and shorter course, sometimes called "baby tamoxifen," was associated with significantly fewer side effects, menopausal and otherwise. It's difficult to determine whether baby tamoxifen has led to increased use of tamoxifen for prevention. Still, more options, for those who are interested, are always better.

RISK-REDUCING SURGERY

The second strategy for reducing the risk of breast cancer involves preemptively removing the tissue that actually forms cancer. This surgery, called risk-reducing or prophylactic mastectomy, can reduce the risk of getting breast cancer to as low as 1 percent. While tamoxifen for prevention can be offered to most women at elevated risk above 20 percent, surgery is primarily offered only to those at the highest risk, such as BRCA mutation carriers. As surgeons, we do not commonly discuss or offer this more aggressive approach to those who are at lower risk levels, such as women with atypia alone or just based on a family history.

THE BREAST HIGH-RISK ADVICE

PATIENTS' WISDOM FOR HIGH-RISK PEOPLE

STACEY SAGER, television reporter: Years ago, I attended a lecture by Mary-Claire King, who discovered BRCA1, and I will always remember it. She said, "I have never met a BRCA person that I haven't found to be extraordinary." I believe this is because we need to take charge of our health at a very young age. I still get emotional about it. It takes an extraordinary person to make such difficult choices and to be really vigilant, not just for the first chapter, but for the decades that follow. That's what it is really about: not making yourself crazy, but being vigilant for your lifetime, even for your children's lifetime, for your family's lifetime. A lifetime of vigilance can save your life. Yes, your doctors save your life, but really you have to get there. You have to get to that point. That's why any of us BRCA women will tell you that we have such a tight community, a sisterhood.

As high-risk people, we've all had family members who took us down that road; we've all watched that path in ways we would rather not. That's what motivated me to always do things differently from how they were done back when my mom had it and things were less developed.

MARLENE C., retired, grandmother: Because of the atypia and my increased risk of breast cancer to 20 percent, I was offered tamoxifen to reduce my risk of getting it. In my particular case, the risk reduction, as it was explained to me, would be dramatic: I could reduce my risk down to less than 10 percent. On the other hand, I didn't feel that my risk was so astronomically high to begin with. I did meet with an oncologist who laid out all the options for me, and we went back and forth. In the end, I decided not to take it. My advice is to get all the information for your case and your particular risk from an expert who really can lay it out for you. It is great to have the option to take a medication to reduce your risk of getting breast cancer, but it's not for everyone.

MY ADVICE

Get genetic testing. If you test negative, but still have other risk factors, consider evaluation with a breast specialist to define your risk and come up with a personalized screening plan. I recommend discussing your risk level with your doctor as early as age 25 to better define it.

Knowledge is power, so use the information from genetic testing and risk assessment to optimize your plan for early detection, or possible medication or even risk-reducing surgery for prevention.

PART II

BREAST CANCER

THE BASICS OF BREAST CANCER

15

WHAT IS BREAST CANCER?

You might have picked up this book because you're interested in learning more about breast health. But what if you are here because you have actually been diagnosed with breast cancer? Before we get into what to do regarding your surgery and treatment options and formulating a plan, I want to remind you of this reassuring statistic: our cure rates are better than ever before, standing at 91 percent or even higher. So, take a deep breath, internalize this number, and allow it to inform your next steps with some very real optimism.

Breast cancer is the most common cancer that a woman can get. You undoubtedly know someone who has already been down this path before you: a family member, a friend, someone from your online moms' group. You are not the first, and you will certainly not be the last. I say this to provide comfort. Given how common it is, there is so much we know about breast cancer, most importantly in terms of how to treat and cure it. I was recently at one of our annual national breast surgical conferences where a speaker showed a slide that stated our collective goal should be zero breast cancer deaths by 2030. As we've seen, we have better technology for early detection and better treatments than ever before. We have moved from film screen mammograms to digital to 3D and contrast-enhanced. Coupled with ultrasounds and MRIs, we have tools to pick up cancers early. On the treatment front, we have better, more effective therapies, with more available each year. It is estimated that more treatment options have been developed in the last decade than in the prior three combined. This is true progress.

Here's the thing: in order for all this progress to take effect, women need to get screened. The same is true for treatments. Opting out of treatment increases breast cancer death rates, in some cases by double. If you are newly diagnosed with breast cancer, keep these factors in mind as you read this and make your own decisions.

Throughout these upcoming sections on the different surgery and treatment options for breast cancer, you'll find that what we call "de-escalation" is a recurring theme. Better, stronger, more effective treatments do improve survival rates for breast cancer, but a lot of research in the last two decades has also focused on determining whether we can safely dial down aspects of treatment without compromising care. Does every patient need surgery? Does every patient need chemotherapy? Does every patient need radiation? Spoiler alert: the answer is no. Now that we have all these options, researchers have started to focus on how we can tailor treatment to patients and give the more aggressive therapies to those who need them while sparing those who don't. Just as breast cancer treatment isn't one-size-fits-all, neither are de-escalation options. While your community of women with breast cancer or survivors can help provide you with invaluable experiential information, you have to remember that their case is not your case, what worked for them may not apply to you.

One of the main factors driving the personalized, individualized care approach to breast cancer is that we have come to understand, starting decades ago, that breast cancer is not just one disease. There are different subtypes and classifications of breast cancer that can determine decision-making about the best path to cure. Finding out that you have been diagnosed with breast cancer can be incredibly scary, even traumatizing. After the initial shock, the key is to take a deep breath, regroup, and start to empower yourself with the facts, which in most cases are sources of optimism.

CANCER SUBTYPES

Let's start with the basics. *Cancer* isn't the only description we use when giving a breast diagnosis. We also have the various kinds of cancer subtypes. The earliest kind of breast cancer that we diagnose is called DCIS, which stands for "ductal carcinoma in situ." The *ductal carcinoma* part means there are cancer cells present. The words *in situ* mean "in their place," which means the cancer cells appear to be contained in their place, within the milk duct. This containment is important because cancer cells that are contained within the duct wall are unlikely to spread. This is why DCIS is called stage 0 breast cancer, and the survival from this type of cancer is estimated to be 99 percent. You will learn more about DCIS in Chapter 22. When the cells break through the duct wall, into the surrounding breast tissue, the cancer is now called infiltrating or invasive breast cancer, which makes up about 70 to 80 percent of all breast cancers. While many of these cancers, particularly those detected early, have not spread, they still have the potential to do so. When cells have spread into the surrounding breast tissue, it means these cells now have access to the pathways for further spreading: blood vessels and lymphatic channels.

Within the category of invasive cancer are different subtypes. The first classification is based on the cell type of cancer that the patient has, with the two main types being invasive ductal and invasive lobular cancers. Together, these two comprise about 90 percent of all new invasive breast cancers diagnosed. Invasive ductal cancers are by far more common, comprising 80% of new diagnoses, with lobular at 10-15%. Of note, invasive lobular cancers can present as vaguer masses where the true extent is more difficult to determine and without specific findings on a mammogram, leading many surgeons to request an MRI for this specific sub-type of cancer. The remaining 10 percent are a combination of other types, like mucinous, tubular, and medullary breast cancer. In general, all of these cancer subtypes, including the rarer ones, are treated mostly the same, with the subtype only describing the cell type of origin. There is also a rare form of aggressive breast cancer called inflammatory breast cancer that involves the whole breast, including the skin of the breast, which looks congested and "inflamed." This rare type, making up only approximately 1 percent of new breast cancer diagnoses, has its own unique treatment pathway. More on this in Chapter 30.

When a biopsy is done and it shows cancer, we then want to learn more about what is or is not making that cancer grow. Tests are run on the biopsy sample to show us the receptors, which tell us more about how a cancer is growing. For invasive breast cancer, we test three receptors. The first two are called estrogen receptor (ER) and progesterone receptor (PR). These two tests help us determine whether or not the cancer is hormone-fed. The third receptor is called HER2/neu, a cancer gene that, while not hormone-based, feeds the cancer in a different way. With these three tests from the biopsy, we can determine the receptor profile. The receptor profile helps us define treatment pathways, including type of treatment, order of treatment, and timing of treatment.

For each cancer diagnosed, the report will tell us both the cell type *and* the receptor profile. Overall, the most common type of breast cancer is invasive ductal cancer that is estrogen/progesterone positive and HER2/neu negative (also referred to as just HER2 negative), with roughly 60 to 70 percent of invasive cancers being hormone positive and HER2 negative. Most cancers are *both* estrogen and progesterone positive; these two receptors are linked. However, approximately 10 to 20 percent of tumors are estrogen positive, but progesterone negative, and a much smaller percentage, less than 5 percent, are estrogen negative but progesterone positive. These different combinations can have implications for treatment, as you will find out later on in Chapter 24. In each case, the degree of hormone positivity is rated and expressed as a percentage, so it's not just "positive" or "negative." A tumor that is strongly fueled by hormones may be 90 or 100 percent hormone positive, whereas a tumor that is weakly fueled by hormones may be 20 percent or even less. These percentages are rated by the pathologist reading the biopsy or surgery specimen, and the results are included in the pathology report you receive.

HER2 positivity is rated on a scale ranging from 0 (negative) to 3+(positive). The 1+ and 2+ results are considered "equivocal." For these two groups a second test, called a FISH (Fluorescence In Situ Hybridization) test is then performed to determine positivity or negativity. The vast majority of the 1+ and 2+ groups on initial testing will end up being negative. So overall, only approximately 20 percent of breast cancers are HER2 positive with either positive or negative hormone receptors; this is the second major subtype.

The third major subtype of breast cancer is called "triple negative," and implies that all three receptors, ER, PR, and HER2/neu, are negative. For these tumors, we do have treatments that work quite well, but, since they are not hormone-fed or HER2/neu stimulated, we know less about what actually drives the growth of these tumors and, therefore, how to treat them in a targeted fashion. The next chapters go through how each subtype of breast cancer impacts our recommendations for how it should be managed.

There are other terms related to a breast cancer diagnosis that you will want to know about. The first, Ki-67, is a measure of how quickly tumor cells appear to be growing and dividing. The second is called lymphovascular invasion or LVI. This term describes whether or not we see tumor cells within blood or lymphatic vessels surrounding the tumor. The third is called tumor grade, which is a measure of how aggressive the cells look under

THE DIFFERENT KINDS OF BREAST CANCER BY CELL TYPE

- Ductal carcinoma in situ
- Invasive ductal carcinoma
- Invasive lobular carcinoma
- Mucinous carcinoma
- Tubular carcinoma
- Medullary carcinoma
- Inflammatory breast cancer
- Paget's disease of the breast
- Malignant phyllodes tumor

THE DIFFERENT KINDS OF BREAST CANCER BASED ON RECEPTORS

- Estrogen/progesterone receptor positive (one or both), HER2/neu negative
- Estrogen/progesterone receptor positive or negative, HER2/neu positive (also called HER2-positive breast cancer)
- Estrogen, progesterone, HER2/neu all negative (also called triple negative)

Note: For DCIS, we usually only characterize estrogen and progesterone receptor status, not HER2/neu. So, DCIS is either hormone positive (90 percent of cases) or negative (10 percent of cases).

the microscope: well differentiated (the least aggressive), moderately differentiated, and poorly differentiated (the most aggressive). These three factors can help define tumor aggressiveness and can independently impact treatment decisions. More often, they are looked at as part of and against the background of many different factors that we consider related to a specific cancer and how to treat it.

A NOTE ON THE STAGES OF BREAST CANCER

When I talk to women newly diagnosed with breast cancer, one of the first questions I often get is, "What stage am I?" And we hear a lot about stages of cancer when people talk about their cancer in the media, or even between friends. Breast cancer starts at stage 0 and spans through stage IV. While a surgeon or oncologist can provide an estimate of stage when you are first diagnosed, *actual* stage is often not determined until after surgery is performed, when we find out precise tumor size and lymph node status. People focus on staging because it provides some information about prognosis, but there are many other important data points in the modern era that provide better and more accurate information for an individual's outcome. We will learn more about these later.

Stage 0 breast cancer is DCIS only. Stage I is typically defined as having a small invasive tumor, less than two centimeters in size, with no lymph node involvement. Stage II includes patients with tumors greater than two centimeters and/or some lymph node spread. Stage III, also known as locally advanced breast cancer, signifies more advanced disease by one of a variety of different criteria. It can look like a large tumor, possibly involving the skin or underlying muscle; extensive lymph node involvement under the arm, under the breastbone (called internal mammary nodes), or above the clavicle (called supraclavicular nodes). In stage III there is no metastasis, meaning the cancer has not spread to other organs. Stage IV disease, however, is defined by spread to other organs, most commonly the lungs, liver, bone, and brain. Stages 0 through III are all treatable and curable. Thankfully, most cancers diagnosed also fall into those stages.

For some patients who have Stage II cancer and many who have Stage III, we will recommend scans of the body to check for spread before any surgery or treatment plan is undertaken. Given the most common sites of spread listed above, the main scans that we order are CT scans of the chest, abdomen and pelvis, which covers the liver and the lungs, along with a bone scan, which is a separate test to look for bone spread. Alternatively, a PET (positron emission tomography) scan can be done, which covers the whole body and all organ systems in one test. This involves an injection of a labeled sugar molecule, which then tracks to sites of high metabolic uptake throughout the body and can show us spots where cancer might have spread. Importantly, none of these tests are completely definitive. A person can have a microscopic amount of disease spread, with scan results that are normal. And the reverse is also true: a scan can show a suspicious finding in the body, but when further testing is done, and sometimes a biopsy of that site is needed, the results are thankfully normal, a "false positive," similar to what was explained with breast imaging. It's also important to know that PET scans are less effective in showing the extent of disease in patients with lobular cancers than the more common ductal cancers. So, the subtype of breast cancer a person has might influence which tests we choose to get.

While most stage IV breast cancer involves recurrence years later from a prior diagnosis, a small percentage of patients, approximately 5 percent, are sadly diagnosed with stage IV breast cancer right at the outset. While stage IV is not considered permanently curable, many women live with it for years and have significant longevity as a result of the extensive number of treatments we have to offer that hold the disease in check. Newer staging systems take a deeper dive into tumor subtypes described above, rather than just incorporating tumor size and lymph node status as the traditional staging system did.

There are often multiple steps in treatment for most women diagnosed with breast cancer. Although, with the exception of stage IV diagnoses, the surgeon's office is the first stop for most women, there are some women who begin other treatments first. The order of treatment is primarily the surgeon's decision. You will learn more about how we decide the order of treatment in Section 6.

THE FIRST CONSULTATION

Regardless of whether surgery is the first treatment or not, the surgeon is usually the first consultation. Surgeons have a lot of responsibility to direct an individual patient's breast cancer journey. I have had the privilege of consulting with tens of thousands of patients over the years, newly diagnosed, seeking guidance on their pathway toward cure. In my practice I meet and establish treatment for approximately ten newly diagnosed women a week. The responsibility to determine an individual's best treatment plan for cure is never taken lightly, and there is no one-size-fits-all. My goal during these first consultations is always to impart the wisdom and experience I have gained over the years to my patients who are about to embark on the path of treatment. My hope is that the information here provides a springboard for you to have similarly meaningful conversations with your own surgeon, oncologist, and other members of your care team.

16

HOW TO BE A "BREAST FRIEND" TO SOMEONE WHO HAS BEEN DIAGNOSED

STACEY GRIFFITH, founding senior master instructor of SoulCycle: My cancer journey would not have gone as smoothly as it did if it wasn't for my friends. I had a friend or two at every single appointment. To be honest, I realize this is a controversial opinion, but I think it's better that you don't totally rely on your partner, wife, or husband—and you rely more on your friends. So it keeps them separate from everything and the partner doesn't have to be the only one to deal with you in that capacity or see you in treatment. Honestly, the tribe, the friends, they're the ones that are holding you up and presenting you back to your partner. You can't have resentment if your partner's not there with you every minute during your journey, sitting with you, crying with you while you get treatment for hours

in New York City in the summer. That's what your friends
are for. My friends were like, "Bring it on, we're gonna make
bracelets, we're gonna look at magazines." We made it like
camp for adults. We had a blast. We got junk food in there.
It was a horrible thing, but I made the best of it, thanks to
my tribe of friends.

DR. PORT: So, maybe it's not you. Maybe you are reading this as
you are about to go through it with a friend or family member
who is newly diagnosed, to support them in their journey, to
arm yourself with knowledge, and to be their "breast friend."
My patients tell me that when you're diagnosed with breast cancer, one of the few silver linings is that you find out just how
many people truly love you. Your family comes to appointments
as a second set of ears. Your friends offer to pick the kids up from
school when you need extra tests that run longer than expected.
Your neighbors drop off dinner after you come home from the
hospital. Your people step up in ways you don't expect. They
walk your dog for you. They empty your drains for you. They
pick up prescriptions for you. They take you shopping for clothes
that make you feel good in your "new body." They laugh with
you. They cry with you. They sit through hours of chemotherapy with you. They send out the updates in the group chat when
you're out of surgery. Simply put, they are there. My patients
often tell me how touched and surprised they are by the people
who go out of their way to lend support. Showing up in times of
stress and duress is one of the most important ways we can show
that we care, and my patients *always* remember who was there
for them. I recently ran into a patient I operated on years ago
while we were both out shopping. She happily launched into an

update on everything she'd been up to, including the wedding of her friend's daughter.

"You remember my friend, Haley?" she said. "She came with me to all my appointments with you!"

"Of course, I remember!" Even if I didn't remember who came to her appointments all those years ago, she sure did.

Being a "breast friend" can be overwhelming. I get that. You want to do your best to support your friend or family member through this difficult time. First off, breathe. Simply showing up is the most important thing. As a friend or family member, one of the roles might be to help with decision-making, which can be complicated when there are multiple options for surgery. More on this to follow. Rest assured you won't be alone; decision-making is where I, the surgeon, come in. It's my job to explain the pros and cons of each option to each individual patient, so she can choose whichever is best for her. The time surrounding a breast cancer diagnosis is one of extreme vulnerability. I see it in my new patients when I am going over options with them, and they often seem to hang on every word. If you are in the orbit of someone going through breast cancer, know that you can help by listening to them as they process their options and supporting them in their decisions. Giving treatment advice can get tricky, so lean into what your person is telling you they want and need. After twenty-five years of supporting patients and their loved ones, here's what I've learned about what actually helps—and what doesn't—when you're part of the support circle.

IF YOU HAVE HAD IT YOURSELF

It's completely natural to want to share your personal experience to help your friend, but it's important to keep in mind that your story might not be the same as hers. While I know it comes from a good place, try avoiding saying things like, "Do what I did." I can't tell you how many times patients who were referred by a friend I took care of have come into my office expecting the same course of treatment because they had "the exact same thing." When I recall the details of the friend's case, no, they did not have the exact same thing! As the surgeon, I must clarify that each case is different and reassure my new patient that I will devise an individualized plan best suited to them.

Focus on your own experience. It might help to lead with "I know no two cases are the same, but here's what I did and why, and here's how it felt for me." In order to avoid regret, women need to be guided to make their *own* decisions. Research shows that the single biggest predictor of regret for patients having a bilateral mastectomy is feeling like they were bullied or pressured into doing a more extensive operation by a family member, friend, partner, or even a doctor. I once treated a fellow doctor, a gynecologist, who opted for a bilateral mastectomy. With every single patient of hers who was diagnosed with breast cancer to follow, she not only "encouraged" them (to put it mildly) to do the same but offered to show her results. I think that most of our mutual patients found it to be a bit much. I know she was just trying to help, but you can imagine how hearing about your gynecologist's bilateral mastectomies and why you should do the same can be a lot.

Additionally, avoid telling a person to ask their doctor about any specific test just because you had it. PET scans, MRIs, and genetic testing, to name a few, are all tests that *some* but not all patients need. Again, share your experience: "I did a PET scan because x, y, z." Then encourage: "There are many tests out there, your doctor will help you determine which ones you need for your specific case."

You'll also want to avoid comparing diagnosis and staging because it can go one of two ways: either it makes the newly diagnosed person feel better or it makes her feel worse. In supporting your loved one, you want to find the balance between acknowledging the very real, very tough mixed feelings that come with any staged cancer diagnosis and fostering positivity.

IF YOU HAVEN'T HAD BREAST CANCER YOURSELF

Not having a personal experience with breast cancer doesn't mean you are not equipped to help your loved one through this journey. Remember that the first part of being a supportive human is simply showing up with kindness and care. On the other hand, given that breast cancer is so common, many women have considered what they *think* they would do if they were diagnosed. Interestingly, this doesn't necessarily translate to what they do when actually faced with it. I can't tell you how many times women come into my office newly diagnosed and tell me they had envisioned a completely different plan for themselves when they walked in. Decisions can change after a deep dive into all the options, which we always do. On top of that, each case is different, and each person is different.

With that in mind, you'll want to avoid sentences that start with, "Well, if it were me, I would . . ." I know you mean well but it's not about you, and your job is to help your loved one determine what is best for *them* regardless of what you envision for yourself. Every week, approximately ten newly diagnosed breast cancer patients walk into my office to discuss their case and plan their care. At least half of them tell me that their friend or family member or colleague or random person they met in Starbucks said, "Well if it were me, I would just take both my breasts off." The truth is, most people, when faced with the options themselves, think through them much more carefully and, when presented with the true risks and benefits, may make decisions different from what they originally thought. One of the most critical aspects of this process is to be confident in your decision-making, so be the friend that helps solidify clarity, not seed doubt.

SOME ADDITIONAL ADVICE FOR ALL OF US

Don't be a swooper. A *swooper* is the term I use for someone who isn't part of any of the decision-making then seemingly swoops in from nowhere on the day of surgery or for the postoperative appointment questioning the chosen course of treatment. Unfortunately, this type of delayed involvement, when a patient has already made her decisions, can be detrimental to her mental processing and throw her for an undesirable loop. I recently called a patient's sister who had flown in from Chicago the night before her sister's surgery. I had not met this sister, and she wasn't included in any of the combined three meetings and phone calls we'd had to discuss surgical options. My intention was simply to let her know everything went well, but it quickly turned into an interrogation. While I understood the sister's desire for clarity, revisiting everything I had discussed previously with the patient and the other family members who were involved wasn't helpful to me or to her sister.

Don't tell scary stories. I know we've all heard horror stories and I know they are scary, but trust me when I say your loved one will benefit more from positivity and the reminder that no two cases are the same than the turmoil that comes with going down the rabbit hole of bad outcomes. Just because your friend's friend developed a second cancer on the other side, doesn't mean that's what will happen in your newly diagnosed loved one's case. Remember that scary stories represent exceptional cases, and overall breast cancer survival rates are above 90 percent.

Walking alongside someone newly diagnosed with breast cancer can be complicated. We're all just trying to do our best. Remember that our patients, *people*, are often in such a vulnerable state that everything they hear gets examined under a microscope. Ultimately, the goal is to be sensitive and thoughtful. To be part of the discussions that happen at appointments, if not as an adviser, then as a second set of ears, a note-taker, a sounding board. When in doubt, just lead with optimism. I always do.

With all that said, whether it is you, the patient, or your friend or family member, this is where your journey begins.

SURGERY

17

LUMPECTOMY VS. MASTECTOMY AND MAKING THE BIG DECISION

ALISON M., marketing associate: Like everyone I know who has been diagnosed with breast cancer—and there are more than a few of us in my immediate circle—I knew I had decisions to make. I didn't really understand the extent to which that was true. At my first consultation, it hit me: I actually had options and would have to make my own decision about which of the different treatment options offered to me to take. The doctor wasn't necessarily going to tell me what to do. It felt empowering, but also daunting, given that this was one of the most important decisions of my life.

I went into the doctor's office convinced I was going to have to get both of my breasts removed for a small cancer in my left breast. Many of my friends had done that and I had this preconceived idea that having a more extensive surgery and removing more tissue had to translate into a better outcome. It made sense in my head.

After meeting with one doctor who thought otherwise and getting a second opinion with Dr. Port, who said the exact same thing, I was reassured that if I wanted to do smaller surgery, a lumpectomy, in my case, it would not at all compromise my chance of survival, which, thankfully, was already quite high. In the end, that's what I decided to do. Don't get me wrong, I would have done *anything* the doctor told me to do to give me the best outcome. Once I realized that the choice was mine, I navigated my conflicting thoughts and feelings by considering what mattered to me. I liked the lumpectomy as an option because it checked all my boxes: I cared about looking good, and I cared about a shorter recovery.

Although I was happy with my choice, I wondered if I was being short-sighted or even selfish for doing the smaller surgery, especially when so many of the people I knew had chosen differently. I found reassurance that, in my case, the outcome would be the same, and that my risk of cancer coming back in that breast was under 5 percent. That was a number and risk I could live with. I know that not everyone can, but I also learned that even with a mastectomy the cancer could come back. Once I knew that, I was happy to avoid a big operation, an overnight stay in the hospital, going home with drains, subsequent procedures to complete my reconstruction, and a greater potential risk for complications. I made my decision, and I never looked back. Of course, once I did that, I discovered that so many other people that I know also chose the same. Even though in the beginning it seemed that so many others had chosen differently. Not that what other people choose should matter. If you have more than one option, my advice is to really consider all the plusses and minuses when weighing your surgery options. There are a lot of plusses with a lumpectomy.

JENNIFER H., homemaker: I was completely on the fence about doing a lumpectomy, which was a reasonable option for me according to my doctor, versus a mastectomy. I listened to everything and heard it all: survival rates were the same, higher complication rate with more extensive surgery. I just knew that I never wanted to have another mammogram again, which, by the way, missed my cancer. Once I went down the mastectomy rabbit hole, I knew I had to do both: if I took off one breast, I still needed screening for the other, which made no sense to me and wouldn't have accomplished my goal in opting for a mastectomy in the first place. I went into surgery extremely nervous. I had no idea what to expect for *myself* despite speaking to so many who had gone through it themselves, my surgeon, and my plastic surgeon. Numerous times. When I woke up in the recovery room, I felt like I had been hit by a truck. By the next day, though, I was up and around, sitting at the table for meals. Once I got home later that day, I got a little better each day. By my third day after surgery, I was completely off pain medication other than extra-strength Tylenol, and I was up and around a lot, taking walks that got progressively longer. By week four I was back in full swing, 90 percent recovered: back in the gym, back in the office, with only small remaining aspects of healing, which continued to improve over the ensuing six months. It was an easier recovery than anticipated and I felt lucky for that. I was so glad to have this behind me, knowing I made the right choice for me. I know I still have the second stage of my reconstruction to get through, but that will be easy compared to what I went through already. The recovery was definitely easier than I expected. I feel lucky for that.

DR. PORT: For the majority of women diagnosed with breast cancer, the first treatment step doctors propose is surgery. So, usually the first and most challenging decision that my patients have to make—that we have to make together—is what type of surgery she should have. There are certainly cases where a patient only has one appropriate option, which will be described below. But most women do have options. Here's how the process goes: I provide the medically safe options, and then she gets to make a choice. The best choice can only be made with the best advice and information, and as usual, that can come from many different places. Importantly, because no two cases are alike, other patients cannot usually effectively weigh in on someone else's options or decision. The right choice for one person may not be the best for another, medically *or* personally. So, while it's helpful to get advice from a variety of sources, be careful to take input from the "chat room," friends, or family members with a grain of salt, as Alison pointed out.

When women come into my office to discuss surgical options, we always end up talking about what their breasts mean to them. Some women are incredibly attached to their breasts as critical parts of their feminine identity. Sensation, sexual function, and appearance all play potential roles in their self-image, confidence, and perception of attractiveness, making the idea of losing one or both breasts potentially devastating. Other women tell me that they've never been particularly attached to their breasts, which makes their loss less upsetting. The role that breastfeeding will play or has played in a woman's life can factor into her feelings about her breasts, her attachment to them, and her decisions regarding surgery. A woman who breastfed might feel that they have "done their job" and no longer serve a purpose. Another may feel a protectiveness toward a body part that served her and her children so well. For many of my young patients, who are either diagnosed with breast cancer before childbearing or considering preventive removal in a high-risk situation, breastfeeding a future child is incredibly important. A mastectomy, complete removal of the milk-producing breast tissue, would eliminate the possibility of breastfeeding from that breast. For some, the inability to breastfeed at all after removing both breasts would be crushing.

You might be surprised to find that a woman's attachment toward her breasts can transcend age. Given that the average woman diagnosed with breast cancer in America is sixty years old, and 75 percent of women who get breast cancer are postmenopausal, I spend a lot of time talking to older women about their hopes and desires not only for survival but for appearance too. I once took care of a seventy-nine-year-old woman who had recently met a new man after decades in what she described as a loveless and contactless marriage and ten years of being single following her divorce. Her new boyfriend of six months, ten years her junior, was totally "into" every part of her, she told me, as she described how he'd brought her the resurrected sex life of her dreams. In particular, he loved her breasts. She was almost eighty years old, but this was, in her exact words, "the crappiest possible time in her life to get breast cancer." Given that she needed a mastectomy, we spent as much time talking about how her breasts would look and feel after reconstructive surgery as we did about her survival rates.

Many of my older mastectomy patients are not so concerned about the nuance of where the incision is or how the nipple will look after reconstruction, but they do focus on how they will look in clothes. As a result, I have developed a minor obsession with "mother of the bride" dresses. Postmenopausal women are usually in their prime years for marrying off their kids; I can't tell you how many women have come to me with their first concern being making it to the wedding. Once I reassure them that they will, we talk about the dress. Not the bride's, theirs. I always want to see a picture of the dress, and I love to reassure them that the dress will look great. After the wedding, it gives me so much gratification to see these women, many of whom have been through major surgery, looking happy *and* stunning. One of my patients showed me a picture of herself at her son's wedding in a spectacular, low-cut, V-neck gown. She looked incredible: beaming with happiness and pride.

"That's not the same dress I remember seeing in the other picture," I said. "You know, the one you showed me before the surgery that you were planning on wearing? What happened?" It turned out, after surgery, she ditched the original, more demure dress and went for the dress she really wanted. She got emotional showing me the picture, saying, "Look how amazing my boobs look! I would never have been able to wear this before my double mastectomy."

Digging deep with patients to get to the bottom of their hopes and fears related to their breasts *before* they go into surgery is just as important to me as that day in the OR when I am standing over them, holding a scalpel. It is a privilege to work in a field where we can have the luxury of factoring aesthetics into decision-making about something as serious as cancer surgery. Every part of surgery makes a difference: where I make my incision to minimize visibility, how long that incision is, how I close the wounds with stitches that go under the skin and dissolve rather than having to be removed. The fact that I do take all of these technical and aesthetic factors into consideration, and discuss them with my patients up front, is often a great source of comfort for them. Seeing their relief when I acknowledge the importance of taking aesthetics into account without ever compromising cancer cure is one of my favorite parts of my job.

Some patients are less concerned about appearance but worry more about functionality after healing. Not of the breasts, but of the surrounding chest wall and the arm. Picking up grandchildren, playing tennis or now pickle ball—the new thing people rave about—and resuming job-related activities are all important considerations when deciding on what type of surgery to have. Many women are able to have relatively easy and straightforward recoveries even after having more extensive surgery, like a mastectomy. While they never go back to the "normal" they had before surgery, there is a "new normal" that they can adapt to and live with. For a small minority, a bigger surgery does translate into tightness across the front chest wall, scar tissue, and even chronic pain. These potential risks of larger surgery must be taken into account by each individual person.

When I talk to women who have the choice of a lumpectomy versus a mastectomy, they always ask "What would you do, Dr. Port?" They want to know if there's one choice that's secretly better than the other. There isn't. As we've seen, people will have many opinions, including the timeless "Just do the big surgery to never think about it again" approach, but the only opinions that truly matter are yours and those of your medical team.

Trust that I've seen many women who claim they always said they would opt for a bigger surgery but, when faced with the choice, opted for the smaller lumpectomy instead. Cut through the noise and listen to the facts and to yourself. The statistics show that 60 to 70 percent of women who can have a lumpectomy do choose that option, appropriately driven by the quick recovery and minimal disfigurement. Many of my patients have a lumpectomy on Thursday, take Friday and the weekend off, and are back at work on Monday.

Regardless of what my patients choose, I always point out to them that everyone's drive behind decision-making is unique, including how *they* feel about their breasts.

LUMPECTOMY

As you make *your* decision, here are the facts. First, it's good to know that there are a number of different names for a lumpectomy; we also call this more limited surgery a wide excision or even a partial mastectomy. So, your surgeon might refer to it as such when discussing this with you. Unless there are extenuating circumstances, a lumpectomy and mastectomy are associated with *equal* rates of survival. I always tell my patients that as long as I can get the cancer out with a clean edge around it, removing more *healthy* breast tissue is not an advantage. Prior to surgery, we often order additional imaging studies or pictures to do everything we can to make sure we know there are no other areas in that breast or the other that are concerning going in. Once that's done, there is no advantage to doing more extensive surgery. While survival rates are exactly the same—and some new data may suggest a tad bit better for lumpectomy—there are a few other key differences:

1. Women who have a lumpectomy usually need a course of radiation to follow. The side effects are typically minimal: mostly some temporary tanning or burning of the skin, which goes away. Going through radiation can be fatiguing. My patients tell me that by the final allotted weeks of treatment, they are pooped at the end of each day. Importantly, unlike chemo, you don't lose your hair with radiation. I know some people are afraid that while radiation can cure cancer, it can also *cause* cancer. Please note that the risk of a secondary cancer related to radiation is far less than 1 percent.

2. Women who have a lumpectomy have a small risk of recurrence in that breast, typically less than 5 percent, which most of my patients find acceptable, especially given that breast cancer can come back or recur even with a mastectomy. Yes, even with a mastectomy, breast cancer can come back in the skin or the pectoralis muscle just beneath the breast. This risk is also thankfully very low, approximately 1 percent. The risk of what is called "local recurrence" is slightly higher with lumpectomy. Many of my patients say, "Well that doesn't make sense: the risk of recurrence with a lumpectomy is a bit higher than mastectomy, but survival is the same. How is that possible?" Well, because patients who have a lumpectomy are usually monitored very closely. *If* a recurrence happens, it's usually picked up early and can be treated again, usually with mastectomy, ensuring excellent cure rates and without compromising overall survival.

3. Lumpectomy surgery is not always "one and done." There is the issue of margins. Cancers don't grow like a ball; they grow more like a star, the tentacles extending into the surrounding tissue. When we perform a lumpectomy, we take the cancer out and try to achieve a healthy "margin" of tissue around

it. Then, we send all of that tissue that we removed to the pathology laboratory for processing and analysis. We sew up our patients after this short surgery and they go home. In the days to follow, we get the pathology report from surgery, which tells us how we did. There, we learn how big the cancer was, whether we got clear edges around it, and, if lymph nodes were removed, if there was any sign of cancer spread. The clear edges or margins are determined by looking under the microscope to make sure the cancer did not come to the edge of what I removed. If it did, I might have unknowingly cut across it, meaning there is still cancer on the other side of that edge, which is unfortunately still in the breast. Getting clear margins happens 80 to 90 percent of the time. If it doesn't, we need to do more surgery, since achieving clear margins is one of the most effective ways of giving a patient the lowest possible risk of recurrence. Needing a second surgery is always disappointing for a patient and her surgeon, but it's worth it when it means being more certain we got as much as we could out. Unlike mastectomy, where the whole breast is removed and there is almost never a need for further surgery, lumpectomy can be a *process*. It's important to know that when choosing this route.

4. After a lumpectomy, women usually do not need any form of reconstruction. The area that is removed usually fills in nicely by itself with fluid, blood, and then scar tissue, leaving little to no defect that can be seen. With some larger lumpectomies, some surgeons do offer a type of reconstruction, rotating tissue internally to fill in a larger space. I encourage you to beware of this approach, which is sometimes pushed but is rarely needed, especially with small tumors. The problem is that tissue rearrangement makes it more challenging to

go back in if the margins aren't clear and further surgery is needed.

5. Importantly—and this is a common misconception we are about to set straight here—having cancer spread to your lymph nodes does not impact decision-making regarding lumpectomy or mastectomy. Women with cancer in their lymph nodes do not need a mastectomy based on that criterion alone.

MASTECTOMY

Here are the main medical reasons why I might, as the surgeon, advise a particular person that in her specific case, a mastectomy might be a safer choice:

1. A large area of cancer, particularly in a smaller breast, may make it more difficult to get all the cancer out, or get a clear margin, and leave someone with a result they are happy with aesthetically. In such cases, a mastectomy may be advised.

2. A person with multiple separate cancers in the breast may be better off with a mastectomy. In selected cases, doing a "double lumpectomy" may be safe, but in general, it is more challenging to get clear margins around two or three cancers, and most data show a slightly higher risk of recurrence. Again, so many individualized factors play into the surgeon and patient decision-making regarding the right procedure for the right person.

3. Having a genetic predisposition to breast cancer, specifically carrying one of the BRCA genes, often means that the risk of

recurrence will be significantly higher, especially in a younger patient. A woman in her thirties or early forties with breast cancer who is found to carry a BRCA gene may be at 50 percent risk for developing a recurrence or new cancer on the other side. Most women in this age group with cancer *and* the gene opt for more extensive surgery, typically double or bilateral mastectomy.

4. For a woman who, for whatever reason, is opposed to receiving radiation or medically should not have it, mastectomy may be a safer choice. This is because, since the risk of recurrence without radiation is substantially higher, the standard of treatment for most women under the age of sixty-five or seventy is lumpectomy *with* radiation, not lumpectomy alone.

Ultimately, when it comes to decision-making about surgery, one of the most critical things I tell patients with *invasive cancer* to keep in mind is that the type of surgery they choose—lumpectomy, mastectomy, or even bilateral mastectomy—will not, in any way, change or influence the type of *medical* treatment they might need to follow. Patients are often very surprised to hear this, and many patients come into my office with misconceptions about this fact. They say, "I think I will just take off both my breasts and then I won't need any chemotherapy or medication to follow, right?" I spend a significant amount of time with patients explaining why this is not the case. As you will see in later sections, the recommendations for medical treatment that may follow surgery can be based on a variety of different factors such as the tumor size, receptor profile, lymph node spread, and other information we get through surgery. Recommendations for chemotherapy or other medications are *not* dependent on the kind or extent of surgery that a woman chooses, and the *exact same* medicine treatment that would be recommended after a lumpectomy would also be recommended after a mastectomy, or even a double mastectomy. I advise my patients to make decisions about surgery independent of what type of medicine treatment (chemotherapy or hormonal therapy) might need to follow, and this is one of the most critically important points for me to communicate to patients up front.

Thirty years ago, I never would have imagined I was about to embark on a career journey that would take me into having this conversation with women every day and help them make one of the most defining decisions of their lives. Now, many years and thousands of operations later, I have seen how time and time again there is no one-size-fits-all. It can be so hard to make this very important decision.

My medical advice to my patients is always that, if they have the choice and are having a hard time deciding, we do the smaller surgery, the lumpectomy, and go from there. Throughout those conversations, though, I always use my standard phrase: "I may be the expert in cancer, but you are the expert in *you*."

18

ONE BREAST OR BOTH?

MARY JOE FERNÁNDEZ, ESPN television commentator, tennis Olympic Gold Medalist, three-time Grand Slam singles finalist: Since my mom was diagnosed with breast cancer, I started getting mammograms at age thirty. I always went into my mammograms a little nervous because of my mom's history but after so many years of being fine, my first reaction was shock when I got diagnosed through a routine mammogram when I was forty-six. They did biopsies and two tumors came back as cancerous. The radiologist was the one who said, "Eh, something doesn't look right here," so I knew it was going to be bad. I froze when I got the call. *Oh my God, it's real*, I thought.

The hardest part of the diagnosis was waiting on the results. What kind of cancer is it? What kind of tumor is it? What does the first biopsy say? What does the post-surgery biopsy say? When you wake up from surgery, you're wondering, *What's going to happen when they dissect the tissue even more? What will they find?* Then you're waiting on other results to decide if you will need chemotherapy. Having something that you can do, that you can control during the waiting periods, helps. For me, it was focusing on getting my mobility back. If I had been told that we had to do both breasts, that it was better for me, I would have done it. Since it was only on the one side, I was comfortable with just doing one, so ultimately that is the choice I made.

I remember that the recovery process was the hardest part for my mom, but it wasn't that bad for me. Being an athlete helped me because I was super disciplined. I did my rehab every day, I did my own exercises every day. Having a purpose and having something to focus on was really helpful.

SARRAH STRIMEL BENTLEY, advocate, motivational speaker, former Broadway actress, and mama of Chance: The option that I chose for surgery was really a tough decision. I was sitting across from Dr. Port and, after being diagnosed and all of my radiology tests were reviewed, we knew the size of the cancer, we knew where it was located, we knew it was in my lymph nodes. She laid out a plan and said, "Here's what our options are, here's what we can do."

She gave me an option for a unilateral mastectomy because my cancer was in my left breast, and it wasn't so small. We had done the genetic testing—which is a whole other ball of wax—but didn't have the results yet. So, I'm sitting there and Dr. Port says, "Listen, we don't know if you have any genetic mutations, but we do know that the rate of success for a unilateral versus a bilateral mastectomy is really the same." You'd think it would be better because you think *Let's take everything off,* but really it's not. We also talked about lumpectomy as Dr. Port walked me through all of my choices. I left her office choosing a unilateral mastectomy.

When I went to go see my plastic surgeon, who was incredible, and we were planning for a unilateral, I started to think about my body. I was thirty-eight years old, a yoga teacher, a professional dancer on Broadway, and I felt like I wanted symmetry, to feel the same on both sides, so it was the same amount of scar tissue. I was still dating, but I had a serious boyfriend. I was like, "Oh God, please don't leave me when I have the cancer." I did not have a ring on it when I was making this decision, which is why I called Dr. Port, and we had a very thoughtful conversation. I still wanted to look good. I remember crying on the corner of Sixty-Ninth and Madison Avenue outside of the store Doggy Style, trying to look for a dog sweater for my dog. I was like, "What am I doing? I'm talking about cutting my breasts off." But at the end of that call, we decided it was the way to go. Then my genetic testing came back, and I did have a mutation and I would have needed to do it anyway. In the end, I'm glad I went with my gut.

JOANNA R., jewelry maker: "If it were me, I'd just take them both off." That's what so many of my friends and family members flippantly said when I was diagnosed with breast cancer. One or two had been through it themselves, but most hadn't. Still, that didn't stop them from being completely comfortable offering up advice. I'd ask them, "If you had colon cancer, what would you do? Take out part of your colon or the whole thing?" Interestingly, many of them would answer, "Well, that depends. I would listen to what the doctor advised and do that." *There you have it.* Breast cancer is one of the only diseases where people other than the patient feel that their opinion counts. Why is that? Maybe because breast cancer is so common and everyone either has it, knows someone who has it, or knows someone who knows someone. Maybe it's because breasts are considered more "expendable" compared to other organs. I would say maybe it's because it's a woman's disease, but these comments often come from other women. The point is everyone has an opinion. But here's the thing: each case is substantively different. My friend Liza had a bilateral mastectomy and told me I should do the same. But she had breast cancer on both sides, which is really rare. I'm not sure she even actually had a choice. I did. Another friend told me that she had "exactly the same thing," but when I dug deeper, I found out that she had biopsies of three different places in her breast that *all* showed cancer. *Not* "the same thing." "Taking them both off" is an option for some women with breast cancer and is even advised in some circumstances, I have come to know. I *needed* a mastectomy on one side. But I wasn't about to impulsively remove the other without weighing all the factors relevant to *my* case and considering carefully the seriousness of the decision. So that's my advice, and my response to the "take them both off" crew.

DR. PORT: When a woman is diagnosed with breast cancer, she often does have options for the surgical approach: lumpectomy or mastectomy, and if it's a mastectomy, single or double. How does she decide? Not *all* women have *all* of these options. Once a woman has decided on mastectomy or been told by her surgeon that this is the safer option for her, the elective removal of the other breast often comes up. When the other breast is completely healthy based on examination and imaging studies, the decision-making is not *medical*. It's personal. Removing a completely healthy breast in no way reduces the risk of cancer coming back in the first breast or provides any medical benefit. So why would someone do it? Three main reasons arise:

1. To reduce the small chance of getting a new breast cancer on the other side. Even after a mastectomy, breast cancer can come back in the skin or muscle on that side. Thankfully, this is very rare; the risk is 1 to 2 percent. Invasive breast cancer can also recur in the body as metastatic or systemic disease. What a specific cancer can't do is *come back* or "migrate" to the other side. Yes, one can develop a completely *new* cancer in the healthy breast on the other side, but that risk is extremely low, ranging from 2 to 5 percent. Removing the other healthy breast provides absolutely no medical benefit related to the current breast cancer recurrence. At least once a week, a woman comes into my office saying, "I just want to take off both my breasts," convinced that removing both will guarantee a cancer-free future. Unfortunately, this is just not the case. Removing the healthy breast reduces that risk by approximately 1 percent, which is quite a low and marginal benefit in terms of added risk reduction. However, for many of my patients who are extremely anxious, even a marginal reduction in new cancer risk is worth taking on bigger surgery. As the surgeon, my perspective is that women who make the decision to do a bilateral mastectomy

need to know what they are and are not accomplishing by opting for bigger surgery. If, God forbid, cancer recurs years after surgery, I would never want a patient to come in saying, "How did this happen?! I thought that by removing both my breasts the cancer could never come back!" It *can* come back, and removing the other healthy breast in no way changes that risk. It's critically important for me to drive this point home to my patients when we are in the thick of decision-making.

2. To avoid future screening. For some women, screening for breast cancer recurrence or a future new cancer can be associated with a lot of anxiety. They just don't want to have the "it's probably nothing" discussion after a mammogram ever again. A woman having a mastectomy on one side would still need to be screened on the other. She may be recommended mammograms, sonograms, and MRIs on a yearly basis after removing one breast. If eliminating the need for future screening is a driver in a patient's surgical decision-making, she might opt for a bilateral mastectomy. I should say that after surgery, some women and their doctors do *feel* things that elicit concern: lumpy scar tissue, small amounts of leftover breast tissue, surgical clips, and sutures. My research from Mount Sinai shows that approximately 20 percent of women come in with concerns about something that they feel, even after a mastectomy or bilateral mastectomy. These concerns often lead to imaging studies like ultrasounds, MRIs, and yes, even biopsies. Patients need to know that while bilateral mastectomies eliminate the need for routine screening, additional imaging and biopsies can still be needed based on things that we feel. So, although bilateral mastectomy may eliminate those yearly screenings, it doesn't

eliminate the need for imaging altogether, and it doesn't always alleviate the anxiety many seek to quell.

3. To achieve a more matched aesthetic result. Breasts—your own, augmented, or reconstructed—are always at least a little bit asymmetric. As someone who examines thousands of pairs of breasts each year, I can say with confidence that very few people have a perfect match. Most women know their bodies well enough to know that too. When a person is having a mastectomy, one breast will look quite different from the other. With no reconstruction, the affected side will be flat with a straight line across. Some of my patients complain that with a larger remaining breast on the other side, there are balance issues and even back pain from being so asymmetric. A reconstructed breast is very likely to look quite different from one's natural, unaffected breast. When the asymmetry will be significant and cannot be adequately corrected by doing a reduction or lift on the other breast, some women choose to remove both breasts for a more matched look. But again, even two reconstructed breasts may not match perfectly! So, if you're opting for a bilateral mastectomy for symmetry and aesthetic purposes, it's critical to have realistic expectations, and to discuss these with the plastic surgeon who will perform your reconstruction.

4. Because of gene positivity, BRCA or one of the others. Just like Sarrah, having a genetic predisposition to getting breast cancer can impact surgery decisions. If you've just been diagnosed, finding out that you carry a genetic predisposition for breast cancer is potentially game-changing information. The numbers and risks vary based on which gene you have and the age at which your first cancer is diagnosed. Many women who do carry a genetic mutation and get breast

cancer, are diagnosed at an early age. And those who are diagnosed with breast cancer young *and* have the gene are at significantly higher risk of developing a new cancer on the other side, potentially as high as 50 percent over the rest of their lifetime. Of course, understanding that this decision in no way affects recurrence rates from the cancer they already have is crucial. But most of my patients who come through cancer surgery and treatment have zero interest in going through it again. With such a high risk of developing a new cancer on the other side, most genetically predisposed women will wisely opt for a double mastectomy.

Sarrah opted for a double mastectomy for aesthetic reasons later to find out her genetic testing results also advocated for that. For many of my patients, especially young ones, one or more of the above factors rings true. However, opting for more extensive or aggressive surgery does have its downsides:

1. More surgery translates to more recovery. Some of my patients think that if you go into surgery for a single mastectomy or a double mastectomy, it's the same coming out. Unfortunately, in surgery, more is more: more recovery, more risk. All the potential risks and complications of surgery, such as bleeding, infection, and anesthesia-related risks, increase with more extensive surgery. For instance, it's possible to have a bilateral mastectomy and develop an infection on the non-cancer side, the side you didn't have to do.

2. Removing both breasts will change how both breasts feel. You may have a loss of sensation or an increase in tightness. Most sensation in the breast originates from the nipple, but even when we save the nipple, we're usually only saving its skin and

rarely preserving full, true sensation. If a woman finds pleasure from the sensory aspect of the breast, preserving the healthy breast can preserve that pleasure; removing the breast will almost always translate to loss. Women may also experience tightness across the front of the chest wall when both breasts are removed, and chronic pain syndromes related to scarring around nerve endings can ensue. Thankfully, in my experience, it's not common for patients to experience chronic, debilitating pain defined as continued pain more than three months after surgery. However, some studies show that this number is 20 percent or higher. It's also, unfortunately, impossible to predict who might experience these dreaded long-term effects.

My patients Laura and Emily had profoundly different experiences after double mastectomies and have shared with me during their yearly follow-up visits over the years since. Here is what they've told me:

LAURA T., life coach: The surgery was really straightforward for me. I was up and moving around the day of surgery. When you came to see me the afternoon afterward, that same day, I was already on my computer sending work emails. You told me, "Don't text on drugs!" But I really wasn't taking any drugs and didn't need any heavy pain medication at all, even in the next few days. I was dreading the surgery, but it was so much easier than I thought it would be.

EMILY M., copywriter: I thought the surgery was about what I expected, in terms of recovery. But over the next few weeks and months, as my healing progressed, I think my scarring did too. Even now, two years later, I have this gnawing sensation on the side of my chest that won't go away. I have seen neurologists, pain specialists, had injections into the site, and have had scans, but there's nothing to see. I even tried acupuncture for it. It's not that it interferes with my day-to-day life so much, but it's always there. Probably a scar formed around a small nerve ending that would be impossible to identify. But it's causing disproportionate pain, like a thorn in an elephant's foot. It's hard to get comfortable sometimes, and the pain has woken me up at night when I turned the wrong way. I knew that the bilateral mastectomy was the only choice for me *before* the surgery, but if I had to do it over again, would I have chosen the same course? Probably. I love that I never have to have another mammogram again, and I definitely have less back pain from my big former breasts. But I'd be lying if I said I didn't have second thoughts sometimes. My advice would be to go into whatever you choose eyes wide open to all the possibilities and outcomes.

Laura and Emily had vastly different short-term and long-term effects, but both their stories emphasize the importance of carefully considering all of your options and going into surgery with full understanding of the benefits, risks, and potential outcomes.

19

SAVE THE NIPPLE?

CHOOSING WHICH TYPE OF MASTECTOMY MAKES SENSE FOR YOU

MELANIE S., chef: I was on the fence about getting a lumpectomy versus a mastectomy. I was given both options. I really wanted to have a mastectomy to get rid of the screening, the small risk of recurrence, and the radiation, which I was told I would need. But the *main* reason I was thinking of doing a lumpectomy is that I didn't want to lose my nipple both for the appearance and the sensation. I didn't want a procedure to reconstruct a nipple that wouldn't look like mine or feel like mine. That wasn't going to cut it for me! When I spoke to Dr. Port about the options, I found out that although many mastectomies involve removal of the nipple, I didn't have to lose mine. My cancer was far from the nipple, so it wasn't a situation where leaving my nipple was going to risk leaving cancer behind and increase my risk of recurrence. Dr. Port told me that after surgery, I would likely not have any significant sensation in the nipple, which was a huge disappointment, but I would get to preserve the appearance. I was willing to make the compromise. I would get the mastectomy but keep my own nipple. So I have both of my own nipples, just one lost its sensitivity. Not exactly what I would call a win-win, but certainly a reasonable option for me.

DR. PORT: Different types of mastectomies have different risks and benefits. As a cancer surgeon, there are a number of different factors I take into account, both oncologic (cancer-related) and aesthetic (appearance) concerns. Typically, the cancer surgeon decides first and foremost what type of mastectomy might be safest from a cancer standpoint, ensuring that to the best of our ability, no cancer is left behind. If reconstruction is elected, the plastic surgeon then weighs in on what options will leave the patient with the best cosmetic outcome, factoring in overall size and shape that you start with; overall size and shape that you desire; nipple size and position; and the length of the scars that will be needed, to name the main considerations. When a woman is having a mastectomy for preventive reasons with no current diagnosis of cancer, we can prioritize aesthetic considerations above all, since there is no concern about leaving cancer behind. Nevertheless, we do prioritize getting all of the breast tissue out. It should go without saying that undergoing mastectomy for prevention and not removing as much breast tissue as possible completely defeats the purpose of having the surgery in the first place! Leaving significant amounts of tissue means leaving risk. And having inadequate surgery is probably a more dangerous situation than having no preventive surgery at all, since it leaves a woman with a false sense of security and she will typically no longer be screened.

Going from most to least extensive, let's review the options for the different types of mastectomies. Of course, not all of them are viable options for all women.

MODIFIED RADICAL MASTECTOMY

A modified radical mastectomy involves complete removal of the breast, including the nipple and areola, along with all of the lymph nodes under the arm. An extensive operation, this is typically reserved for patients with the most advanced disease—often including significant lymph node involvement—where we need to completely remove all of the breast and lymph nodes to get rid of all the cancer. Women can certainly still have reconstruction with this type of surgery and achieve a good cosmetic result, but there is a risk of lymphedema, a temporary or permanent swelling of the arm, related to removal of all the lymph nodes. That's why this procedure is reserved for those who need it because of disease spread to the nearby underarm nodes called axillary nodes.

SIMPLE OR TOTAL MASTECTOMY

Don't get confused by different names used for the same thing. Just like an ultrasound is the same thing as a sonogram, and a lumpectomy is the same thing as a wide local excision or even a partial mastectomy, there are a few names for the same type of mastectomy. Despite what the names may suggest, a simple mastectomy is the exact same thing as a total mastectomy. Both refer to a removal of all breast tissue, including the nipple and surrounding areola. It's similar to a modified radical mastectomy, minus the node removal.

SKIN-SPARING MASTECTOMY

A skin-sparing mastectomy is just like a total, simple, or modified radical mastectomy from the breast tissue removal standpoint. The only difference is the implication that we will save as much of the skin envelope as possible to optimize aesthetics and reconstruction. Most women with early-stage disease can have this type of mastectomy, and it's what we favor for them, as long as the cancer is far enough from the skin that we don't have to worry about leaving it behind on the skin flap. A skin-sparing mastectomy always implies that reconstruction will be done at the same time, since the saving of more skin is done with the specific aim of creating space for an implant or tissue taken from another part of the body. With any of the mastectomy versions we've reviewed so far, the nipple and areola are not preserved. We remove them and leave a seam, a straight line across the chest or reconstructed breast mound where the nipple used to be. In a separate procedure later on, the nipple can be reconstructed over that scar, through tattooing or building up a nipple projection from the scar tissue. The decision to reconstruct the nipple mound, tattoo it, or do nothing at all is very personal. Many women claim that they spent their pre-mastectomy lives worrying about covering up their nipples and don't see why they would rebuild them to deal with the same problem? Other women want their appearance to be as close as possible to what they had before. The decision of what to do about restoring nipple appearance is completely individualized. It's an aesthetic decision, and a personal one, not a medical decision, as all of these are equally safe.

NIPPLE-SPARING MASTECTOMY

Perhaps one of the biggest advances in improved aesthetics for breast surgery is the increased utilization of nipple-sparing mastectomy, which involves preserving the nipple areolar complex, and essentially the entire skin envelope of the breast. Twenty-five years ago, when I was first starting out, performing nipple-sparing mastectomies was much less common. There was significant skepticism about the safety of the surgery, and lack of clarity regarding who was or was not an appropriate candidate. I have been privileged to be part of the generation of surgeons who have helped to develop and grow the utilization of this technique. Now, in some major breast cancer centers, nipple-sparing mastectomies comprise up to 50 percent of all mastectomies performed. We can make an incision in the crease of skin at the lower border of the breast, where it's barely seen, or in a straight line radiating out from the nipple, and remove all breast tissue through that incision. The benefit of this approach is predominantly aesthetic, which, of course, can have profound positive implications on the psychological impact of having a mastectomy. The preservation of one's own nipples can, first, eliminate a step in reconstruction as described above, and, second, provide much comfort by preserving yet another aspect of your original appearance.

Importantly, most nipple-sparing mastectomies do *not* preserve the sensation of the nipple. There are newer techniques to painstakingly dissect out nerves to the nipple and preserve them or place nerve grafts under the nipple with the hope of the nerves regenerating and providing sensation. Nerve grafts can be very expensive and are not always reimbursed by insurance. The jury is still out on the effectiveness of these approaches, and there is some concern that preserving nerves results in preserving more breast tissue and a less-than-optimal mastectomy, in terms of removing all of the tissue. We need more long-term data to give us a better idea of the true effectiveness in sensation preservation, but it's important to know that these techniques are being used and championed for those who feel that this is a priority.

A NOTE ON DRAINS

With almost all mastectomies, drains will be placed at the time of surgery to absorb and expel the fluid the body makes as part of the healing process. We place them in the wound site, and they come out through the skin, usually with a bulb attached on suction to collect the fluid. Patients often go home from surgery or the hospital stay with these drains in place, which need to be emptied. Many patients can do this themselves, or they enlist a willing, non-queasy family member to assist. In the first few days after surgery, that fluid can appear blood tinged, but as healing progresses, it gradually gets more yellow or straw-colored in appearance. We ask our patients to keep track of the amount of drainage fluid they empty each day. As the healing process settles down, the fluid production tapers off, and the drains can be mostly painlessly removed in a quick office visit.

20

TO RECONSTRUCT OR NOT TO RECONSTRUCT?

HILARY P., child behavior specialist: When I learned that I had breast cancer, I decided to have a lumpectomy. I thought it would be straightforward to do a forty-five-minute surgery. It would be in and out, and I'd move on. Unfortunately, we found out a week after the surgery that my margins were not clear, and Dr. Port had to go back again. Again, the margins were positive. Neither of us had any idea that there would be so much more cancer. From my mammogram and MRI, the cancer looked very limited. As I found out, these tests are unfortunately not perfect for determining how much cancer is there. In the end, I had to have a mastectomy to get it all out. Clearly, this was not my first choice. To be honest, I was pretty devastated. Luckily, reconstruction was an option for me, as it is for most women. It helped a lot to know that I would have *something* there after surgery and look reasonably normal in clothes. I only did one breast. I didn't want to do more to my breasts than was necessary; that's why I picked the lumpectomy in the first place. Thankfully, the plastic surgeon did a really nice job of making my reconstruction look like the other breast. I had a flap of tissue from my belly moved and reattached to build my breast. I swear, they look nearly identical. When I go to get a mammogram, the tech usually has to ask which side had the mastectomy. Plus, the belly procedure—which was a big deal, I'm not going to lie— meant that I got a silver lining tummy tuck. Whether you are having a mastectomy because you chose to or you need to, be sure to consider all your options for reconstruction, and pick what's best for you even if the surgery is longer and more involved. For me it was worth it.

AMANDA KATZ, certified clinical medical assistant: I decided not to do reconstruction after my bilateral mastectomy. I don't know if it has anything to do with the fact that I'm a lesbian, I am married, I have my children, and I was never attached to my breasts as a feminine definition. Breasts, to me, were just a part of my anatomy; it wasn't because I have breasts that I identified as a woman. That is one thing that is very different from many women. Also, I'm more of a masculine, tomboyish kind of woman. The way I dressed never accentuated that part of my body. That's one side of the story.

The second side of the story is that I had been working at a plastic surgeon's office for eleven years by the time I got diagnosed with breast cancer. Over the course of that time, I had seen the number of women increase steadily that are being diagnosed with breast cancer, or with BRCA or any kind of genetic mutation. Most of them would come to our office for implants. I've watched these women go through the long process that it takes from getting your diagnosis to your new breasts. You have to get the cancer removed, the breasts removed, and then procedure after procedure of expansion, implants, the second surgery. Sometimes we need to do a little bit of revision and then you do your nipples and then you tattoo, if you want to do that part. It just seemed like a lengthy process that I had no desire to go through. As a forty-eight-year-old woman when I got diagnosed, I didn't want to constantly deal with implants. Watching others go through it for over a decade, I always thought, *What would I do in that situation?* It happens to one in ten women. It's a very large number. I thought, *One of us working in this office is probably going to get it.* It happened to be me, thank God, because I feel that mentally I was able to handle it, especially since I didn't feel any kind of attachment to my breasts.

The recovery process is its own thing. You come home and you have these drains in your chest, which are uncomfortable and irritating. They tug at the side of your body, and you want to be careful that they don't spill. Not having anything there felt okay at first. I wasn't conscious of it. I had a bra on with some compression underneath. The bulkiness of the bra made it look like I still had breast tissue in there. It wasn't until the first time I was allowed to take a shower when I took the bra off and looked down and saw nothing that I almost passed out. It was such a shock. You're used to seeing breasts for over thirty years and suddenly it's just bruised and flat. I remember sitting on the end of the bed and my wife asking, "Are you going to pass out? Are you going to be okay?" I was crying and thinking, *What the hell just hap-*

pened? It had all happened so quickly. The whole thing from diagnosis to surgery was boom, boom, boom, and boom, now I'm here. A few days after seeing my post-surgery chest I realized, *Oh yeah, this really happened.* It was almost like a dream. Crazy.

I didn't realize that going flat was something that's becoming a movement. It's funny how a lot of women tell me, "I wish I would have done what you did." They're much older, in their sixties, seventies, and patients of ours. I don't want to discourage anybody because everybody has their own choice, but it seems like putting yourself through that reconstruction process is harder than having breast cancer in some ways. You have to worry about implants. Things happen to them, they rupture, they descend, they move in their pockets. They have to be maintained and cared for, and I don't have time for that. I want to focus on other things.

If you're feminine and love your breasts and you need them to be there, God bless you, put them back. The silver lining for you is that you can have any size breasts you want. But if you're on the fence, going flat is the way to be for me. It's so much lighter on your body. If you're an active person, there's nothing in your way. Imagine if you're a surfer, imagine there being nothing between that surfboard and you. Personally, I love it. Smaller is better, and in my case, none is the best.

DR. PORT: Across America, approximately 60 percent of women who have mastectomies undergo some type of reconstruction, which is usually performed at the same time as the breast surgery, called immediate reconstruction. The statistics can vary widely based on location and access to plastic surgeons with an expertise in reconstructive techniques. The Women's Health and Cancer Rights Act (WHCRA) passed in 1998 is a federal law that provides for insurance coverage of nearly all procedures related to breast reconstruction after mastectomy. In some states, like New York, laws have been passed to ensure that patients are at least offered reconstruction at the time of their breast surgery or referred elsewhere if those services are not available. These laws include ancillary procedures done to achieve symmetry with the other breast if only unilateral mastectomy is performed. Once you have decided to have a mastectomy with reconstruction, your breast surgeon will often refer you to one or more plastic surgeons that they work with for consultation. Most major academic and busy community medical centers across the country that have specialized breast centers also have specialists in breast reconstruction who are skilled at and offer the full gamut of reconstructive options. You can also check on the American Board of Plastic Surgeons website where you can verify board certification in plastic surgery, which usually signifies a high level of expertise and training.

It's important to know that reconstruction can happen immediately in the same surgery, as it did with Hilary, or it can happen in steps, with only the first step happening at the same time as the main mastectomy procedure that the breast surgeon performs. The different potential processes for reconstruction are explained below. There are also many who do not undergo reconstruction based on the increased medical risk from a longer procedure or the multiple procedures that reconstruction entails, advanced age, which is often associated with increased risk, or, like Amanda, based on personal preference. The "flat after mastectomy" movement has gained significant popularity, even among young women.

FLAP RECONSTRUCTION

There are two ways to reconstruct a breast shape and size: with implants or with tissue from another part of the body, also called autologous tissue. There is significant variability on the type of reconstruction offered, in part based on expertise available in the region where you live. The most common type of autologous tissue flap is called a DIEP flap. The DIEP (deep inferior epigastric perforator) is a blood vessel that runs in the lower belly wall. When the tissue is removed and reattached to vessels in the chest by the sternum or breastbone, it allows for continued blood flow to the flap. We can only perform this surgery for women with enough lower belly volume to create one or two new breasts, depending on whether she is having a unilateral or bilateral mastectomy, so it's often not an option for women who have relatively little abdominal belly fat. It can also be tricky for women with prior scars on the belly from a C-section or another surgical procedure that might have cut the blood supply. The DIEP flap surgery can take anywhere from six to ten hours, requires a few nights' hospital stay, and means recovery from *both* breast and abdominal surgery and wounds, which can be longer and more arduous. With a big incision on the lower abdomen, getting mobilized to walk and move around takes longer than for women who have implants and have no abdominal wall surgery. By contrast, women who have mastectomies without reconstruction or choose implant-based reconstruction are often up and about a few hours after surgery. At Mount Sinai, we keep all of our patients overnight after any mastectomy to monitor for bleeding or other complications, but patients who have implants or opt out of reconstruction are quite mobile and go home within a day of surgery. As you consider your options for reconstruction, keep in mind that being in reasonably good physical condition and able to tolerate the risks of a longer, more involved surgery is definitely important for getting a flap. There are also flaps that can be taken from other parts of the body, when the belly is unusable. There are thigh flaps, and tissue can also be taken from the buttocks. These procedures are less commonly performed.

While almost any plastic surgeon can do implants, the flap surgery entails painstakingly cutting and reattaching small vessels to each other, which requires special training in microvascular surgery. Not all plastic surgeons have this advanced skill, which often requires training beyond the typical six years of plastic surgical residency. As discussed previously, your breast surgeon should guide you to the right plastic surgeons who they work with regularly, typically within the same hospital system. And your breast surgeon should also factor in what type of reconstruction you may want based on a discussion up front. When patients convey to me that are interested in hearing about all the reconstructive options and seem like good candidates for both, I send them only to plastic surgeons who have the expertise to perform both implants and flaps so they can get the full picture and decide for themselves. So, as Hilary advised, make sure you know all of your options for reconstruction based on not only your body type but the services available to you where you will be treated.

IMPLANTS

Reconstruction with implants is another form of reconstruction that is very commonly performed and can also happen at the same time as the mastectomy. However, implant reconstruction can also involve different stages or steps, with only the first step happening with the mastectomy. With implant reconstruction, the empty space where the breast tissue was usually gets filled by saline balloons called tissue expanders. The process of placing these temporary tissue expanders (or spacers, as some people call them) allows the remaining skin to heal without being under pressure from larger implants. Usually a few weeks after surgery, when wound healing is further along, the plastic surgeon will inject saline into the spacers through a small port under the skin as a quick procedure that can be easily done in an office. This expands the spacer and gently stretches the overlying skin. The extent of skin stretching and expander injections over the weeks and months following surgery depends in part on how large the implants will ultimately be; the more we stretch the overlying skin, the more space we have to accommodate implants. At the same time, we usually have a conversation about the desired size and shape of implants with the patient and the plastic surgeon as to the aesthetic goals and what is possible. A woman who wants to be quite a bit bigger than she was before her mastectomy might require additional expansion and skin stretching to make more space for a larger implant.

Once the expansion is complete, the tissue expanders get exchanged for permanent implants in a smaller surgery performed by the plastic surgeon who usually goes in through the same mastectomy incision; this does not require drains or an overnight hospital stay. A lot of patients ask if it's possible to have a one-stage, direct-to-implant reconstruction. In many cases, it is. If a person doesn't want much bigger breasts than they had and can have a nipple-sparing mastectomy, the skin envelope stays essentially the same. Placing an implant the same size as the original breast may be feasible without stressing the skin or compromising healing. Please be wary of the surgeon who promises a "one-and-done" reconstruction no matter what. Until the breast surgeon and the plastic surgeon are *in there*, they don't really know how the skin will look after surgery. Everyone's blood supply is different. Plus, young women have a more robust blood supply than older women. If you *are* promised direct to implant, there is a chance that the breast surgeon will leave more breast tissue behind to create a "cushion" for the implant so it doesn't press directly on the remaining skin, which can lead to an increase in recurrence risk. This is counterintuitive as the whole point of a mastectomy is to decrease risk. Of course, anyone can develop a cancer recurrence after a mastectomy, no matter how thorough; all breast surgeons have seen that happen to their patients, including me. But leaving more breast tissue behind definitely increases that risk. I have seen examples of it in patients who have had less-than-thorough mastectomies elsewhere with extensive amounts of residual breast tissue left behind, either inadvertently due to suboptimal technique, or intentionally to achieve a potentially better cosmetic result and to be able to go direct to implant. I call these "boob job mastectomies" and they should not be done. Not for aesthetics, not for expediency, not based on patient demand, not based on doctor recommendation. Remember, a mastectomy is not a boob job. A mastectomy is performed to cure or prevent cancer.

Now let's talk about the implants themselves. One of the biggest advances in breast cancer reconstruction has been the advent of different types of implants. More shapes allow for more patient choice in how reconstructed breasts look or feel. For example, silicone may feel softer and cause less rippling that can be seen through the skin. Over the years, the topic of implants and their composition has also been fraught with controversy. In 1992, despite widespread use for both reconstruction and aesthetic augmentation, silicone implants were taken off the market due to thousands of claims of adverse health effects. The complaints included everything from leakage to autoimmune disease to "silicone-related illnesses." The company that manufactured the implants, Dow Corning, filed for bankruptcy, and only saline implants were used in the United States for the next fourteen years. In 2006, after data showed no real association between implants and thought-to-be related illness, silicone implants were determined to be safe and reintroduced, which restored more options for women—with and without breast cancer.

Subsequently, another form of implants started to be used on a trial basis before becoming more widespread. This was a new variety of silicone implants called textured implants. Textured implants have a rough surface that was thought to reduce tightness and contracture while adhering to the surrounding tissue and preventing the implant from rotating and flipping, as many smooth implants did. Textured implants had been used in Europe since the 1980s, so they were by no means new. In the late 1990s, a report arose of lymphoma, a type of lymph cancer, developing around a breast implant. This new entity was named breast implant–associated anaplastic large cell lymphoma or BIA-ALCL. The FDA recognized the potential link between breast implants and BIA-ALCL in 2011, and reports of cases coming in involved textured implants almost exclusively. Now here's the rub: given the widespread use of implants, it is impossible to determine or even estimate the risk of BIA-ALCL. Estimates range from 1 case in every 3,000 women to 1 in 30,000; either way, far less than a 1 percent risk. Nevertheless, implant-related illnesses were back in the headlines. Textured implants were banned in

2019, and remaining inventory was recalled. As someone who had literally thousands of patients with textured implants, this was a confusing, chaotic time. The calls did not stop. Were their implants safe? Should they have them switched to non-textured ones? Did the word *recall* mean they had to be removed from their bodies? It didn't, I told them; just that implants not yet used should be returned unused to the company. The risk was less than a fraction of 1 percent; I didn't think it was appropriate to tell women they had to change out their textured implants. After all, when the recall happened, the FDA recommendations were to be aware of symptoms of BIA-ALCL (swelling, fluid, pain, redness at the site of the implant) but not proactively remove or replace implants in all cases. The cases of BIA-ALCL are few and far between, and, most importantly, it is *curable* by removal of the implant and the surrounding capsule with no need for more advanced treatments. Many of my patients with textured implants did electively switch their implants to non-textured ones in an abundance of caution. No one knows for sure why textured implants seemed to be associated with BIA-ALCL. Was there bacteria hiding in the nooks and crannies of the textured surface? Did the surface trigger an inflammatory response, which then spiraled out of control into a full-blown lymphoma? Due to the rarity of cases, we will probably never know, and textured implants are no longer in use in America as well as most of Europe.

Given that certain types of breast implants have been associated with discontinuation or recall, it's no wonder that there are a lot of myths floating around about health issues related to implants. Here's the bottom line: implants—or any form of reconstruction, for that matter—do not increase the risk of breast cancer. It doesn't matter if they're for augmentation or reconstruction after mastectomy. They also do not hide cancer recurrence or make it more difficult to detect on physical examination after a mastectomy or on screening tests, such as mammogram and sonogram.

FLAT

Reconstruction encompasses some of the greatest advancements in breast cancer care in the last decade. But that doesn't mean everyone wants it. As we've seen, an increasing number of women are opting for no reconstruction, citing reasons including quicker recovery, shorter surgery, fewer additional surgical procedures, and, of course, not feeling that reconstruction is of any added value to them. Some women may have an aversion to having a foreign body or implant. The reality is that implants, in their various forms, have gone through more than one round of being taken off the market due to safety concerns. No matter how safe the data say that implants are, many women still have concerns. This isn't surprising, especially given so much mixed messaging over the years. And with the DIEP flap procedure being a long, arduous surgery that not all women can have, many opt out of reconstruction altogether.

Increasingly, patients are asking for plastic surgeons to get involved with their wound closures, even when no reconstruction is being performed. While breast surgeons are more than capable of closing a mastectomy incision—and I pride myself on my wound closures and making them as nice as I can—there are definitely cases where the plastic surgeon has significant expertise to offer, especially when women with large breasts want to go completely flat. There is often redundant skin in these cases, that if not perfectly trimmed will create folding and not lie completely flat against the chest wall. A few years ago, a patient came in and used the acronym AFC (aesthetic flat closure), which I had never heard before, and asked me if I could work with a plastic surgeon to perform it. Now, AFC is a widely embraced concept. If a patient asks, I am happy to have a plastic surgeon help to get her as close as possible to what she wants.

Reconstruction or no reconstruction, flap or implant, saving the nipple or not. There are so many decisions to make once a woman decides that she will be having a mastectomy. It can seem overwhelming at first, but that's why I work with every woman I see to weigh all the options to choose the optimal course of action for her. Mastectomy surgery is extensive and can potentially impact a woman in so many ways: physically, psychologically, sexually, she won't go back to the normal she once knew, but ideally, she'll find a new normal, one that she is happy and satisfied with.

LYMPH NODE SURGERY

IS IT ALWAYS NECESSARY AND, IF SO, HOW MUCH?

MARIANNE J., retired police officer, NYPD, and proud dog mom: I was diagnosed with the most common kind of breast cancer, invasive ductal cancer. It came as a surprise, but not out of nowhere. I was sixty-three years old, around the average age that women get diagnosed with breast cancer. I had no family history, which, I have come to learn, is the case for most women diagnosed. Because I had been getting yearly mammograms, my cancer was small, about one centimeter. I knew that there would be lots of decisions to make. Here's what I didn't realize would be a decision: What to do about my lymph nodes, to check or not to check? Let me explain. A lot of my friends who had been diagnosed with breast cancer and had the same kind of cancer that I had got a few of their lymph nodes under the arm on the same side as the cancer removed to check for spread. As they were told, their treatment would change a lot if it had. Imagine my surprise when my surgeon told me there was new information that has come out recently to show that *long-term* survival and recurrence rates were the same for women who got their lymph nodes checked as it was for those who didn't, and that treatment was mostly the same either way. I was shocked. My doctor told me the smaller a cancer is, the lower the likelihood of nodes being involved. My chance was less than 10 percent. I felt great about that until I wondered what would happen if I was in that 10 percent. Taking out the majority of the cancer surgically was the most important part of my treatment plan, but it would not be the *only* treatment I would get. In the majority of cases, those additional treatments are very effective at "mopping up" any microscopic spread that might be left behind, including spread to my lymph nodes. Dr. Port examined me and pressed under my arm. She didn't feel any enlarged nodes. She reviewed my mammogram and ordered a sonogram to make sure they didn't look abnormal on those tests either. When they came back as normal, I was comfortable not checking nodes during my surgery. I am a "less-is-more" kind of person, so if I could safely avoid even a small procedure, I was happy to do so.

DR. PORT: Lymph nodes are also called glands. These glands live all over our bodies, grouped together like bunches of grapes, as part of the body's immune system. If you've ever had a sore throat and felt tender, swollen glands in your neck, you've felt lymph nodes overfilled with immune cells to fight the nearby infection. Lymph nodes are part of our bodies' filtration system, stopping infection—and yes, cancer—from spreading further. In addition to the lymph nodes in your neck, there is a grouping of twenty to forty lymph nodes that live under your arms, in your armpits. We call them axillary nodes. When *invasive* breast cancer tries to spread, the lymph nodes under the arm on that side of the body can be one of the first stops. Lymph node status, meaning whether or to what extent cancer has spread to the lymph nodes under the arm, has traditionally been a really important part of cancer staging and treatment. If lymph nodes show cancer, a patient is often considered stage II or even stage III, if many nodes are involved. It is certainly important to take nodes out if they show a significant amount of cancer spread, but the good news is that with screening, we are finding smaller and earlier cancers, which have lower likelihoods of spread. Early diagnosis saves lives while allowing us to do less aggressive surgery. To be clear, smaller cancers can also be aggressive and spread, even when they appear early, but that's the exception, not the rule.

The evolution of lymph node surgery for breast cancer over the last thirty years is a prime example of the progress we've made in dialing down the aggressiveness of treatment on all fronts, including surgery. Changes to lymph node surgery have come as three major advances. Before the mid-1990s, when cancers were often larger and more advanced at the time of diagnosis, they frequently involved nodes and there was only one option for lymph node surgery: taking them all out. This procedure, called an axillary dissection, involved making a large incision under the arm and removing *all* nodes, an important and often necessary step toward getting all the cancer out. Especially since we didn't have as many other effective treatment options back then, surgical removal of all the cancer, including in the nodes, was our best bet for cure. When the nodes were removed, we'd send them to the laboratory where they would be analyzed, generating a report on how many nodes were actually removed and how many contained cancer. The result would be reported as a fraction: 3/25 nodes had cancer, for example. Sometimes, we'd see that 0/30 nodes had cancer, meaning all thirty nodes that were removed were normal, which was great news for the cancer spread but bad news for the patient's recovery from a larger operation that hadn't been necessary in the first place.

Axillary dissections are associated with both short- and long-term potential risk. They dramatically affect postoperative course and recovery and can create sensations of tightness, pulling, or numbness under the arm. They can even affect the arm's range of motion. Perhaps the most well-known risk is the dreaded complication of lymphedema, or swelling of the arm, which can be temporary or permanent. It's not totally predictable who might develop lymphedema. The risk after an axillary dissection is anywhere from 20 to 40 percent, and no one would want to live with it long-term. For patients, especially those who work with their hands, functionality tends to be *the* primary and overriding concern, above aesthetics or anything else. I've treated a violinist, a cosmetologist, an electrician, a fellow surgeon, and an FBI agent. All of them with cancer that had spread to lymph nodes under their arm and all of them concerned not only about how their breasts would look after surgery, but how their upper bodies would function and how surgery might affect their performance, in and out of their jobs. The conversations we had were understandably intense. The FBI agent told me she would rather die than lose the ability to shoot and protect herself and her partner. As someone who works with her hands, I could relate to my patients in an extremely personal way. I'd always wonder what I would do on the receiving end. *No one* wants to compromise her cancer cure, but quality of life—which encompasses aesthetics *and* ability—is a huge concern for patients that needs to be prioritized. Thankfully, axillary dissections have long been abandoned as the only way to check lymph nodes under the arm.

The first lymph node breakthrough: in the mid-1990s, surgeons developed a new way to check lymph node status without having to remove them all. Enter the sentinel node biopsy, a more limited procedure to remove a selected, smaller number of nodes. Rather than removing them all, we could now choose which nodes to remove during either a lumpectomy or a mastectomy by injecting dye into the breast with cancer. We use a radioactive dye that emits a signal and a blue dye that is visually detected. Many surgeons, including myself, use both of these at the same time while some are comfortable using only one or the other. The dyes travel to a few lymph nodes under the arm in a pattern of drainage that mimics the way cancer might travel. Dye travel does not mean that cancer has spread, but it shows us where the cancer would spread first. Once we know where to look, we are able to make a small incision under the arm and cherry-pick the node or few nodes where the dye traveled. If cancer was seen in that small sample, we could always take more, but because at this point in time we were finding cancers smaller and earlier, nodes were more often normal, showing no signs of spread. With normal or negative lymph nodes, no further surgery was needed. Plus, the risks of sentinel node biopsy were significantly lower than an axillary dissection on all fronts, with a risk of lymphedema, numbness, and other surgical side effects at less than 5 percent. Naturally, this was a huge step forward for our patients, and sentinel node biopsy became part of the standard of care for invasive cancer surgery in the late 1990s.

I watched this exciting new technique, sentinel node biopsy, become widely adopted during the later years of my surgical training and early years of becoming an attending surgeon myself. The first sentinel node biopsy case I scrubbed in on was in 1998, in my final year of surgical training. I was eight months pregnant. I was so big I could barely get myself close enough to the OR table to assist the attending surgeon performing the case. I remember we briefly paused and worried if the radioactive dye we injected into the patient would affect my baby, still in my belly, which was now smushed against the patient on the OR table?! No one knew for sure, and we went ahead. Fortunately, we now know that the answer was definitely "no." My colleagues and I went on to participate in and develop a number of research studies demonstrating the safety and accuracy of the sentinel node procedure in a variety of different scenarios, expanding its use further and further, and more widely replacing axillary dissection.

The second lymph node breakthrough: after its adoption, sentinel node biopsy *only* was the standard of care for patients who were found to have normal or negative lymph nodes. Many patients who were found to have cancer in their lymph nodes were still undergoing the axillary dissection. During sentinel node biopsies, we would often send the lymph nodes to the laboratory for a very quick read called a frozen section. If lymph nodes were found to have cancer, we would take the rest out during that same surgery. Other times, we would get results from surgery a week or two later, and if even one node had a little bit of cancer, we would return to the OR for an axillary dissection. Ironically, taking out more lymph nodes rarely showed any additional disease. Surgeons started to wonder if it could be safe to skip the axillary dissection, even in those with positive nodes, sparing yet another group of patients from the bigger surgery. Then, in 2011, a landmark study called the Z11 trial revolutionized the practice of lymph node surgery again. In the trial, women who underwent lumpectomy (mastectomy patients were not eligible) with one or two positive nodes were randomly sorted into two groups: one to receive no further surgery and the other to have a complete axillary dissection. In follow-up, the rates of recurrence were just as low in patients who had no further surgery and the rates of survival were just as high as those who went on to have complete axillary dissection. This was big news. It meant that even patients with one or two positive nodes could safely be spared the major surgery and its potential long-term implications. Interestingly, the patients who had more surgery were found to have additional positive nodes in 27 percent of cases. All patients got additional treatment, including radiation with their lumpectomies, along with some getting chemotherapy and antihormonal therapy. These additional treatments "mopped up," as Marianne explained, any disease in the nodes that might have been left behind for those who did not have additional surgery, equalizing the outcomes. When trial results were first published, we were able to reassure yet another group of patients, those with one or two positive nodes who were having lumpectomies, that no further surgery was needed.

Shortly after, a comparable trial called AMAROS showed that radiation without axillary dissection was equally as effective for patients who had mastectomy and positive nodes. As a result of these two studies and others to follow, we surgeons have become increasingly comfortable with telling our patients with low numbers of positive nodes that they can safely avoid axillary dissections. This has translated into less aggressive surgery options for my violinists, my waitresses, my fellow surgeons, and frankly just more women, all of whom would love to avoid a higher risk of lymphedema if possible.

The third lymph node breakthrough: in 2023, yet another landmark study from Europe, called the SOUND trial, was published. This one took previous research a step further, asking whether there was a group of patients with such a low risk of having positive nodes that we could avoid even the sentinel node biopsy. In other words, if nodes were so likely to be normal, did we even need to check them? In this trial, mostly older patients with small, favorable cancers with low propensity for spread underwent an ultrasound targeted to the lymph nodes under the arm and, of course, a physical examination. If nodes were felt to be normal on exam, and appeared normal on ultrasound, patients were randomly assigned to either undergo sentinel node biopsies or no lymph node surgery at all. Just like Z11, the SOUND trial researchers found that at five years of follow-up, recurrence and survival rates were the same for both groups. For those who had lymph node surgery, 14 percent were found to have some cancer in their lymph nodes. Assuming the same was true for the group who did not have any lymph node surgery, and approximately 14 percent also had positive nodes in this group, it was clear that additional treatment that was given took care of almost all of this disease. Only 1.6 percent of these patients had a recurrence in their lymph nodes from disease left behind, compared to 1.7 percent in the group that had nodes removed. This was the trial that allowed me to safely offer Marianne the option of omitting the sentinel node biopsy, which she chose to do.

Thanks to these three major advancements, women with

breast cancer have more options and decisions to navigate than ever before. They may need all of their lymph nodes checked and removed, some, or none at all. As always, there is no one-size-fits-all. A friend who needed all of her nodes removed may have had significant spread. A cousin who had sentinel node biopsy may have a more aggressive tumor that was suspicious for spread. With all of the well-meaning advice you may get, it can be overwhelming to navigate all of the information. I hope the guidelines below will help you arm yourself with questions for your doctor about the best course of action *for your case* and why.

When you have invasive cancer, these are the general situations when we can omit lymph node staging with sentinel node biopsy, when we usually still recommend it, and when we should still perform an axillary dissection, taking them all out.

1. **When it might be okay not to take any lymph nodes out.** In the SOUND study, women with small (less than two centimeters) cancers that were of the most favorable profile (estrogen and progesterone positive, and HER2/neu negative) with no other ominous features had a low likelihood of lymph node involvement and could omit the sentinel node biopsy procedure. Nevertheless, for younger, premenopausal women the information obtained from lymph node staging can often direct subsequent treatment. Many surgeons tend to still advocate for checking lymph nodes in our younger population (below the age of sixty is the cutoff some use, while others use premenopausal versus postmenopausal status), where discovery of lymph node involvement might change recommendations. As further information with long-term follow-up emerges, the decision-making process to check or not check lymph nodes at the time of surgery will continue to evolve.

2. **When we still want to check.** If you have one of the more aggressive types of breast cancer, triple negative or HER2/neu positive, the status of the lymph nodes is extremely important in dictating treatment. For larger tumors of all types, the chances of spread are higher, and therefore checking lymph nodes and removing the ones with significant amounts of cancer spread is still beneficial. Again, younger patients, for whom lymph node involvement might dramatically impact treatment decisions, may be recommended to undergo sentinel node biopsy as part of their surgery.

3. **When we still want to take them all out.** Patients who are found to have lymph node involvement before surgery, by doing a biopsy of suspicious lymph nodes that we can feel or see on imaging studies, fall into a separate group. This is sometimes an indication to do chemotherapy first, with the goal of containing and controlling disease, and shrinking everything down (more on this later in Chapter 25). For patients who undergo surgery first, when nodes are involved, we often do remove the entire group of nodes. Of course, this larger surgery can increase the risk of lymphedema. In addition, with patients who have cancer in their lymph nodes at diagnosis who get chemotherapy first, often the lymph nodes will shrink down, and the cancer disappears. When this happens at surgery, if testing shows that the cancer has completely melted away, we can avoid taking them all out. Alternatively, if there is still residual cancer present, we will often proceed at that point and remove them all. There are ongoing, exciting clinical trials to tell us more about the need for complete lymph node removal, a procedure that most of us, and our patients, would prefer to avoid, unless it might

compromise cancer cure. For those with positive lymph nodes, a procedure called TAD, targeted axillary dissection, can sometimes be performed to remove a more limited number of nodes, rather than all, if the amount of lymph involvement also appears to be limited.

These impressive and continued developments in lymph node surgery have led breast surgeons to safely do less, not more. In cancer treatment, we call that de-escalation: dialing down our interventions and our therapies to incur fewer side effects while achieving the same excellent cure rates. As a surgeon, it is extremely exciting to safely offer many of my patients less surgery. Figuring out if de-escalation is appropriate requires a high level of expertise and expert medical guidance. Keep this in mind as you navigate your path and integrate stories from your community with information from your doctor on your particular case.

22

TREATING DCIS

UNDERSTANDING THE EARLIEST FORM OF BREAST CANCER

JUANA D., therapist: I was doing my mammograms every year. When they found a dot in my left breast, my doctor called it a calcification and sent me for additional screening. I did another mammogram with close-up views, and then a biopsy. The biopsy showed something called DCIS, the earliest form of breast cancer. I cried. A lot. Then I realized that, even though no one was going to tell me I was lucky after getting diagnosed with breast cancer, I was. Not for having cancer but for the cancer I had, compared to what I could have had, or what it might have turned into.

I had a lumpectomy, followed by some radiation. I tried tamoxifen, but it didn't agree with me, so I stopped after six months. While it would have lowered my risk of cancer coming back, since my risk was only about 5 percent, I was good with that.

It's five years later, and yes, I still stress out before every yearly mammogram and visit to the doctor. I had one false alarm two years ago; my mammogram found something in my other breast, and I freaked out. It was so demoralizing, but the biopsy came back negative. I was so grateful. Now, I barely think about having cancer, except for a few days before mammogram time. Once it's done, I'm good. The survival rate for the type of breast cancer I had is 99 percent. It won't be what takes me off the planet.

DR. PORT: Even though invasive cancer is the most common kind of breast cancer, 20 percent of all newly diagnosed cancers are noninvasive. These cancers are called DCIS, which stands for "ductal carcinoma in situ." The term *in situ* means "in its place," so DCIS means there are cancer cells present, but they're still contained within the duct structure of the breast tissue. This containment is critical, since cancer cells confined to the duct cannot really spread and would have to break through the duct wall to become invasive in order to even have the potential for spread. When a person like Juana has a biopsy that shows DCIS, she gets diagnosed with stage 0 breast cancer, which is extremely treatable. As a reminder, and because many patients have asked me about this, a biopsy can't spread cancer, or "push it" outside the duct, or make it invasive.

Most DCIS cases are diagnosed through mammograms, which pick up calcifications. They show up on mammograms as little white speckles or dots, like grains of salt. Many calcifications seen on mammograms are normal; they can come from scar tissue or even normal cell changes. Approximately 20 percent of new or increased calcifications on a mammogram are cancer, and often found to be DCIS. We can also diagnose DCIS as a mass a person can feel or as changes to the nipple like scabbing or new discharge, especially if it's bloody.

Even when it comes to DCIS, different patients have different options. The various types and grades of DCIS can have a significant impact on treatment recommendations and decisions. DCIS is grouped into different gradations: low, intermediate, and high. High grade is the type riskiest for progression and spread, compared to low and intermediate grades. For DCIS, we also check estrogen and progesterone receptors, but not HER2/neu. DCIS is almost 90 percent hormone-based, as opposed to only 60 to 70 percent for invasive cancers. The remaining 10 percent of cases are often more aggressive and higher grade.

The standard of care treatment for DCIS is surgery first, so your first stop will be a breast surgeon's office. From there, the options for surgery include lumpectomy and mastectomy, and the decision process can depend on many of the same factors as for invasive cancer.

LUMPECTOMY

Lumpectomy for DCIS is exactly the same as it is for invasive cancer. We make an incision in the skin and remove a small portion of the breast. The goal is to remove the mainstay of the cancerous area and to get clear margins or edges all the way around it. Typically for DCIS, we want to see a minimum two millimeters of clear tissue all the way around. After a lumpectomy, most women do need some additional treatment. Again, as with invasive cancer, this can be a course of radiation. Sometimes we give patients antihormonal therapy too. Since 90 percent of DCIS is hormonally driven, antihormonal therapy can be quite effective in reducing the risk of recurrence. Some women have both radiation and medication to give them the lowest possible risk of recurrence. Lumpectomy alone is not usually standard of care for most patients given that it is associated with a double or more risk of recurrence, compared to with radiation. Historically, trials of lumpectomy without radiation for DCIS demonstrated a range of recurrence of 50 to 70 percent higher when radiation was not given. The grade of DCIS, the amount, and the margins all influence risk of recurrence without radiation and account for the large range in recurrence rates seen across various trials. For example, in one trial studying the omission of radiation in patients with low-risk DCIS, recurrence rates without radiation approached 20 percent at 20 years. For higher risk disease, it was closer to 30 percent. Radiation reduces these risks of recurrence by half or more, across almost all groups. Critically important is to know that about half of the time, when DCIS recurs, it comes back as something worse: invasive cancer. Our bottom line is minimizing that risk. We don't want to undertreat someone today, only to deal with a worse problem down the line.

MASTECTOMY

Similarly, a mastectomy for DCIS is the exact same surgery we perform for invasive cancer. We use the same incisions and the same decision process for saving the nipple based on proximity to the cancer. It may sound extreme to perform such a large operation for such early-stage disease. As with invasive cancer, the size of the area involved with DCIS can impact plans for surgery. Ironically, DCIS can involve more of the breast than invasive cancer, since it can spread along the ducts across a broad area without ever becoming invasive, making it harder to do a successful lumpectomy. If the goal of surgery is to get all of the cancer out with clear margins, that might not be achievable with a lumpectomy. DCIS might also be multicentric, or involve multiple separate areas of the breast. Doing a double lumpectomy is more challenging and may be associated with a higher risk of recurrence because of more extensive disease and also the challenge of effectively radiating two spots instead of one. Finally, since DCIS develops in the milk ducts, which all converge on the nipple, it can get very close to the nipple and even extend into it. When the nipple is involved, a lumpectomy is often not possible. There is a type of lumpectomy, called a central quadrantectomy, that removes the whole nipple and areola along with the ducts directly underneath, but the breast often looks disfigured without the nipple, so mastectomy can be a better option aesthetically. Patients might also opt for a mastectomy if they find out they have a genetic predisposition to breast cancer and want to prevent the higher risk of it coming back in that breast or the other. This is especially true for women who are diagnosed young, which so many patients with genetic mutations are. In the end, personal preference might also be a driver for mastectomy. With mastectomy for DCIS, the operation is the cure. There is no radiation to follow. There is no medication recommended. It's true, breast cancer can still come back after a mastectomy, but that risk is 1 percent, and no additional treatment would reduce it to zero. So, a mastectomy gets a woman the lowest possible risk for recurrence, with no further screening, and sometimes, women just want to be done with treatment after surgery, once and for all.

THE ROLE OF CHECKING LYMPH NODES IN DCIS

We discussed that lymph nodes are often checked for spread in patients with *invasive* cancer. Thankfully, DCIS has almost no potential for spread to the lymph nodes or anywhere else. When we can do a lumpectomy for DCIS, we have no reason to perform sentinel node biopsy and check for spread unless we suspect that the cancer might actually be invasive. Remember that a needle biopsy only removes small snippets of tissue. When the area of disease is so small, the needle biopsy may remove most or even all of the cancer, but the opposite can also be true. In about 10 percent of cases, when a needle biopsy shows DCIS, when we remove the entire area through surgery, we can find out through analysis of the surrounding tissue that the cancer is "upgraded" to invasive. And it's not like a plane trip where an upgrade to first class is a good thing. In our world, an upgrade is when the biopsy shows one thing, and then, when the area is removed, we find something worse. Here are some red flags for the possibility of invasive disease in DCIS patients: When women come in with an actual lump that they feel, instead of just calcifications on a mammogram, finding invasive cancer is more likely, and a surgeon may advise checking lymph nodes. There are also times when the pathology report from the biopsy will say "cannot rule out invasion or invasive cancer." This essentially means that the biopsy showed DCIS, but we're not sure if there could be invasive cancer there too. In situations like those, your surgeon might advocate for adding sentinel node biopsy to your lumpectomy.

When we perform a mastectomy for DCIS, however, we often perform sentinel node biopsy. Taking a few lymph nodes out with a mastectomy does not significantly impact recovery time or ability to use the arm in any way following surgery. More importantly, if we skip a sentinel node biopsy and find invasive cancer later on, there's not much we can do without the breast, where we inject the dye to find the right nodes. With lumpectomy, the majority of the breast is still in place, and we can always perform sentinel node biopsy to follow at a second operation, if needed, based on the findings.

23

IS BREAST CANCER OVERTREATMENT A THING?

WHEN MIGHT YOU CHOOSE THE "WATCH AND WAIT" APPROACH?

ARIEL Z., IT consultant: I was diagnosed at fifty-five years old. My mammogram picked up my cancer as a small area of calcifications. My biopsy showed intermediate grade DCIS. I met with two surgeons who explained that this was stage 0 breast cancer: not the kind of cancer that typically spreads. Even if it did, they told me, it would probably take a long time. We went over all the options, but a lot of the treatment options sounded extreme to me. Radiation? Mastectomy? It seemed like a lot for a cancer I'd just been told I barely needed to worry about. I had a question: What would happen if I did nothing? To my surprise, I wasn't met with a blank stare. Instead, my doctor told me that she sometimes offers what she called the "watching and waiting" approach for people

who don't have invasive cancer, which could be a reasonable option, especially in a cancer like mine, intermediate
grade, where no one really knows the course. I was told
that some, but not all, DCIS does progress on to invasive cancer but that I would be monitored closely. In the
meantime, I am taking an antihormonal medication to try
to stave off my cancer cells' further growth and spread.
I know that this might not be the right course for everyone, but, honestly, I am so happy that I don't have to have
a mastectomy or radiation or anything else that seemed
like overkill for what I have. I was so grateful that my doctors offered me options.

DR. PORT: While breast cancer is a life-threatening disease, and there are still approximately 40,000 breast cancer–related deaths each year, we have always known that some breast cancers are slow growing. They can percolate for years without becoming life-threatening at all. With technology getting better and better and cancers being detected earlier and earlier, do *all* cancers, even the ones we pick up as microscopic specks on a mammogram, go on to become life-threatening and require treatment? Can some cancers go untreated without ever affecting overall health or survival? Autopsy studies suggest approximately 10 percent of women who die of other causes are found to have had breast cancer inciden-tally. One to two percent of these cases are invasive and the rest are mostly DCIS. The statistic is analogous to prostate cancer in men, where the incidence of previously undiagnosed prostate can-cer found on autopsy ranges from 20 to 60 percent for men over the age of eighty. The watchful waiting approach for prostate can-cer has been an option for selected cases since the 1980s and 1990s, mostly because the treatment for prostate cancer can be quite onerous. Prostatectomy is big surgery. Radiation, another option for some men, is also quite aggressive. Both treatment options are fraught with significant side effects that profoundly impact quality of life, like potential loss of sexual function or continence.

While watchful waiting can be a preferred option, it does have limitations. We may not fully know what we are dealing with without any surgery at all. As you learned in the last chapter, even when a biopsy clearly shows DCIS, surgery—either lumpectomy or mastectomy—might reveal that there is actually invasive cancer present, known as an upgrade. So, watching and waiting on DCIS may end up being watching and waiting on invasive cancer. So what ends up happening in these cases where there is something worse? Follow-up exams that might happen three to six months later should reveal changes that, in turn, prompt intervention. If there is something worse there, the close follow-up will often reveal signs of progression. Then action can be taken, with very little risk that some delay in treatment will significantly impact progression of disease, but it is a risk. I have had patients who have ended up regretting their choice for watchful waiting or delay in treatment of DCIS as a result of having invasive cancer ultimately found at surgery.

The second limitation is that we have not yet figured out a fail-safe way to identify *which* patients can be safely watched. The major decision factors for taking a watchful waiting approach are:

1. **Age:** Being diagnosed with DCIS at age forty is different from being diagnosed at age eighty. If an eighty-year-old woman is diagnosed with DCIS and doesn't treat it, she most likely *will* go on to die of other causes given that the average current

lifespan for a woman is eighty-two or thereabouts. Can a woman with DCIS at age forty go another forty years without *any* intervention? The long-term plan for patients in this age group is not entirely clear and has yet to be established.

2. **High-grade disease:** As discussed, high-grade disease is a harbinger of potential for invasive disease or progression to invasive disease. Nevertheless, clinical trials are underway to help determine scientifically how watchful waiting on DCIS plays out, even for those with high-grade disease. If we can peel back and de-escalate treatment with invasive cancer (more on this soon), why wouldn't we do that with DCIS?

In summary, the theme of de-escalation has led to significant progress in eliminating the overtreatment of breast cancer on many fronts, including surgery. Research, currently ongoing, will continue to lead us toward more precise approaches, dialing down treatment when we can, particularly as it relates to DCIS, the earliest breast cancer.

THE BREAST SURGERY ADVICE

ADVICE FROM MY PATIENTS ON
SURGERY DECISIONS

GINA S., media professional: For young women, you get a lot of unsolicited advice from other people who think they have your best interests at heart and want to help. Even long after my surgery, when most of my treatment was done, people, total strangers even, were still giving me advice. I was on jury duty, and I went to get lunch on our break. I sat at the bar at some restaurant, and I was chatting with the woman next to me. I don't remember how it came up, but at some point, I started telling my cancer story. "Darling, you really should think about getting a mastectomy in the next few years, it's gonna come back," she said. "I appreciate that. Thank you so much," I replied, maybe being too gracious. I felt at peace with what she said because I knew I wasn't going to do it. I believe in listening to my body and my doctors and my providers. I'm not using this outside influence of someone who hasn't gotten through it, but it was weird. Anyway, expect a lot of unsolicited advice about your surgery choices, especially if you are young.

SARRAH STRIMEL BENTLEY, advocate, motivational speaker, former Broadway actress, and mama of Chance: What's right for me isn't gonna be right for another person. Depending on your build and how much you have to work with, reconstruction can look different for everyone. I got under-the-muscle implants and I love them. They look fabulous, but they're not perfect and that's okay.

I have a group of friends that were my friends before, and we all got breast cancer. We'll sit around the dinner table and I'll be like, "My nipples are crooked," then my friend will say, "Well, I don't have any." It's different for everyone. I decided to keep my nipples because there was no cancer near them, but my nipples are sisters, not twins. Probably because I had to have radiation on the cancer side, which affected that result. If you get a double mastectomy and you cannot keep your nipples, the silver lining is that you can get them tattooed beautifully. I know an incredible tattoo artist. They're all over the country and can make them very symmetrical.

The nipples are there, and I have scars that I did not get lasered. I wear shirts where you can see a little of my scars on the side, and I rock them because they remind me of the journey I went through. It reminds me of everything I've done to get here. I think they're beautiful, just like my hard nipples that are slightly crooked that you can see that I don't wear a bra with. I feel comfortable and okay with that because, at the end of this journey, the woman that I've become through this fire is so in her body. There is such ownership of this power that I'm like, "Look at my crooked nipples and my scars, y'all, because I'll tell you a story." I'm not perfect and no one is, and I think there's something beautiful about that.

HANNAH STORM, sports anchor, producer, and director:
I found my recovery from the lumpectomy to be really seamless. My breasts looked a little different, but they eventually popped back into form. The pain was not bad at all. I went back to work within one week and could have gone back sooner.

MY ADVICE

Lumpectomy (often followed by radiation, but not always) and mastectomy are both options for the surgical treatment of breast cancer, and for most women newly diagnosed with breast cancer, it's the first critical decision they will have to make. This decision-making process is the pinnacle of moving past a one-size-fits-all approach in cancer care. While delaying cancer surgery for long periods of time (more than a month) is never advisable, getting the cancer "out" is not a medical emergency either. This is where it is important to take a beat. Make decisions that are best for yourself—in collaboration with your inner circle and your doctor—for the long term. As I have said many times before here, the survival rates of breast cancer are higher than ever before. So this is often a decision for the duration of a long life ahead.

TREATMENTS BEYOND SURGERY

CHEMOTHERAPY

LONGER AND STRONGER VS. DIALING IT DOWN

This is the part I find patients dread and focus on the most when newly diagnosed. They ask questions like, "Do you think I will need chemotherapy? If so, how much and how long? Is the treatment intense? Will I lose my hair?" These and other questions are addressed in the following pages. While I am not a medical oncologist, the type of doctor who makes decisions about chemotherapy and other medical treatments, I have helped send many people down this path. As you will see in the stories below, there are so many different possibilities. A word to the wise: while my patients directly below all received chemotherapy, increasingly, as you will learn, many can avoid it, due to newer tests to help us decide and other new therapy options.

STACEY SAGER, television reporter: I generally find among people I have spoken to, those who do surgery first and then oncology, most feel that the surgeons are the most rosy. So, you know, you have your surgical experience, and it goes really well, and hopefully, you feel like, *Wow, I really aced this.* Then you meet with your oncologist and that meeting really hits you—and I've talked to a lot of people with different cancers about this. I know that now more than ever before, many people, including people I've spoken to, no longer need chemotherapy based on newer studies. Regardless, the meeting with the oncologist hits you and hits you hard. If I was going to give people advice, I would say, buckle up for that meeting because it becomes clear in that meeting that it's really a marathon. It's not a sprint. The surgery might've been the sprint and you might've won that battle, but there may be a longer battle still to get through. This is not to knock oncologists, because they're fabulous; this is to say that they are going to present you with their version of what lies ahead. That version is longer and possibly tougher than anticipated.

SARRAH STRIMEL BENTLEY, advocate, motivational speaker, former Broadway actress, and mama of Chance: As a person who lives in her body and works with her body, I was really worried about going into treatment. My body was my calling card in every way. I was terrified that this treatment could make me gain weight. I had heard that was a possibility. I say *heard* because I try not to look online; it's like all bad Yelp reviews. Don't try to forecast what you think is going to happen to you, because everybody's journey is different; you're just going to cause yourself more anxiety.

But yes, one of the big things that I was terrified about was gaining weight, because all I heard was with chemotherapy, you're going to get fat. Then because of the following treatments to keep the estrogen out of your body, well, that's going to put you in menopause and that's probably also going to make you gain weight. There are all these horror stories about gaining twenty to twenty-five pounds at the end of treatment and everybody's online saying this, that, and the other thing. Turns out I didn't gain or really lose weight.

Movement really does keep you in your body when it's changing. And it's proven to help treatment be more successful and ease the side effects. So, move your body. I walked the dog on my sickest days. I would come in from walking the dog, take all my clothes off in the hallway, and just crawl in bed, but I did it, and I felt better.

VIVIANA FIGUEROA, wife, runner, guardian to twins: I started running in 2014, when I was in my thirties. I had never run before and wasn't into sports as a child. One of my coworkers was running the New York City Marathon and I went to watch. I spectated at mile twenty and saw a lot of people walking. At the time, I was going through some deep relationship issues with a partner and got motivated to try it. What's the worst that could happen? You have to walk.

I signed up for some 5Ks and 4-Milers and eventually set out to run my first New York City Marathon in 2015. Through the running journey, I joined a running group, and that's where I met my husband. I attended a lot of group runs, started dating, and then eventually got married to him, five years after I started running.

I went big from the very beginning and went for the

marathon. Eventually I learned how to swim, got back on my bike, which I actually found more therapeutic than running. I read this post one time that said, "You can't be sad and ride a bicycle." That was true. I found myself smiling a lot when I was riding a bike.

When I got diagnosed with breast cancer, running through treatment wasn't a priority. It became that much more therapeutic when I could get some miles in, but I wouldn't call it running. I dialed it back to just walking, but I'm a nerd for numbers. I was tracking my distance, tracking my pace, even with walks.

Chemotherapy side effects were really strong. There were days that I could not make it out the door just to the CVS that was right next door to my apartment building. So, half a mile or a mile was a big win during treatment. I took a break during treatment in between chemotherapy and surgery and was able to get some longer distances, like a three-mile walk, nice and slow, just enjoying the fresh air. It's interesting because I became a running coach and became that much more interested in numbers and heart rate and pace and distance. I remember talking with my therapist about running and how I was feeling about my loss of fitness levels at the moment. Her advice to me was "How about you just go outside without your watch? Just go outside and look at the trees and enjoy the scenery." It was liberating because without looking at the numbers, I wasn't reminded of how far I had fallen back. I took another break in between surgery and radiation and once radiation was fully over, in October, I told myself, "I'm running the marathon next year."

During recovery, I started running thirty seconds at a

time: run thirty seconds, walk thirty seconds, and I'd go out for thirty minutes at a time. As far as I got was as far as I got. Prior to being diagnosed, during the marathon, I would run past the hospital, Mount Sinai, at mile twenty-three. There's an unofficial name to that mile because of the hill: Ninja Hill. When you're surrounded by runners talking about the marathon, everyone talks about Ninja Hill, mile twenty-three. "Be careful. It's going to sneak up on you. It's really long."

Ironically, when I got diagnosed, I found myself circling around looking for parking on Ninja Hill and often thinking, "Wow, am I ever going to run Ninja Hill again?" I would park on 106th street and walk over to the hospital feeling the incline and thinking, "I don't know if I'm ever going to run this again."

The day of my first marathon after recovery, I didn't know that the doctors and nurses came out. I started in the final wave and by the time I got to mile twenty-three, the spectators had dwindled down, but there happened to be at least fifteen doctors and nurses all cheering in front of Mount Sinai. I stopped to tell them I just beat cancer and they started jumping up and down. Some years the crowd has been bigger than others, but I always, always manage to stop. That has now become the most grateful mile of the marathon.

I can't help but reminisce on how many times I had to park my car and walk into Mount Sinai to sit in a chemo chair that I didn't want to sit in. Now I'm running past it. It's become that much more special. I've now run ten marathons, and this next one is going to be my fourth post-cancer.

JILL MARTIN, *TODAY* show contributor, entrepreneur: I needed chemotherapy. I did four dose-dense red devils and I did four Taxols instead of the twelve Taxols; I was allowed to pick. *Let's see if my body can take it*, I thought. And it took it. Every time I showed up, my blood count came back okay and I was able to receive treatment. I got through it.

I believe that two things can be true at once. You could want to save your life and want to look and feel like yourself and feel pretty and want long hair. That is a life lesson that opposing feelings can be held at the same time, and don't let anybody tell you otherwise. When people said to me, "Well, thank God, you're alive," I was like, "That's a very low bar." I strive for the extraordinary.

During chemo, I did cold capping to save my hair. Cold capping is excruciatingly hard, but it pays off on the back end. When I would complain about how hard it was, Dr. Port said, "Sit in the [expletive] chair and put on the hat." I appreciated that because that's what I heard in my head. My hair was my identity. Everyone was like "You are the girl with the gorgeous blond hair." I wanted to hold on to that. That's who I was. I'm not that anymore. I still have gorgeous blond hair; it's just different.

DR. PORT: For patients who have surgery first, treatment from there usually follows this order: chemotherapy (if needed), then radiation (if needed), and finally antihormonal therapy (if needed). That's why you will see the following chapters in that order. However, as I hope you are beginning to understand, not all women need all of these types of treatment. So, if a specific treatment is not required for an individual, she moves on to the next treatment that is needed *in her case*. Most patients with invasive cancer, and even some with DCIS, are recommended to have some additional medical treatment after surgery. When many of my patients hear that, they say something along the lines of, "I don't understand. If you take the cancer out—why am I not done? Why do I *need* additional treatment at all?"

The answer to this very important question is that while cancers grow over what is usually months to years leading up to one's diagnosis, a cancer cell or cells can flick off from the main tumor. They can get into the lymphatics, the veins, or the arteries that run through the breast tissue, which are superhighways to the rest of the body, and escape. Notably and frustratingly, there is no test to determine if this has happened: no blood test, no body scan, no symptom that would allow us to know if this has happened. These cells can circulate and ultimately land in other organs in the rest of the body—lungs, liver, bone, and brain are the most common sites. When this happens, we have what is called metastatic disease, also called stage IV disease, which is a more severe situation. Stages 0 through III of breast cancer are *all* curable. Stage IV is not. That's why we want to prevent it. Clearly, surgery can't go everywhere in the body. As much as I wish we could, a surgeon can't find and pluck out individual cells circulating around and get rid of them, the way that I can precisely excise a mass in the breast or lymph nodes under the arm. Unlike a surgeon's hands, medicine *can* travel throughout the entire body. Whether we take them by mouth, through a patch on the skin, or through an IV, all medications get absorbed into the bloodstream and circulate.

One of the biggest misconceptions about treatment and progress is that it always means longer and stronger: new and improved treatments, more aggressive treatments, more treatment options. We *do* have treatments for breast cancer—surgery, radiation, chemotherapy, and medicine—that work extremely well and play a potential role in treatment for the individual patient. It's also true that in the past decade, there were more effective therapies for breast cancer developed than in the previous thirty years combined. That means there are exponentially more treatment options than ever before. But that's only part of the story. Over the past two decades, just as much research has been devoted to exploring how and when we can actually *pull back* on some aggressive treatments and provide equally effective results. Sure, we could give everyone everything. We could throw the kitchen sink of all treatments at every single patient with breast cancer. However, a large number of women would unnecessarily suffer the side effects that can even be more serious than a breast cancer diagnosis (leukemia, a type of blood cancer, is one very rare potential side effect) related to extreme treatments they potentially don't even need. When I started to practice, anyone who had a tumor larger than one centimeter—which is pretty small—was considered for chemotherapy, even if their lymph nodes were negative. If your lymph nodes were positive, you definitely got chemotherapy. That meant that almost all women with invasive breast cancer with anything more than the smallest tumor had chemotherapy. Today, we use personalized, precision medicine to treat women as individuals, and in some cases, chemotherapy is no longer needed.

To break it down, chemotherapy is medicine usually given in IV form as a drip over hours. From there, it circulates, going everywhere in the body—bloodstream, organs, lymph nodes, and, yes, even the breast—and it can tackle and kill cancer cells wherever it goes.

For many patients, chemotherapy is something to be avoided, if at all possible, predominantly because of side effects. The potential side effects include feeling poorly, lower blood counts, immunosuppression, hair loss, weight gain, and even secondary cancers that are worse than breast cancer—but again, please note that this last one is rare, with less than 1 percent of patients affected. While it might seem the most trivial, hair loss is extremely traumatizing for many patients. For some, having to wear a wig or being bald means a loss of privacy about their diagnosis. It makes them immediately recognizable as a cancer patient in their community, in the workplace, and to everyone they encounter. The other part of chemotherapy that patients dread is how long it takes. Chemotherapy is usually broken into a series of treatments called cycles given every week or so over the course of months. The side effects are no joke, but at the end of the day, chemotherapy is one of our most effective weapons to cure breast cancer and prevent recurrence, metastatic disease or spread, and death.

In our field, we have been asking for years how it might be possible to scale back on this toxic but potentially lifesaving aspect of breast cancer care and still deliver the same survival rates and excellent outcomes. Many breast cancer researchers have made it their life's work to explore if it would be possible to avoid these severe treatments and have reasonable certainty in selected patients that we could achieve the same results. They wanted to know if we could move past the "scorched earth, one-size-fits-all" approach. Could we figure out who does and does not need specific treatments and deliver more personalized care? Here's the answer: we already have. As with radiation therapy, we have been able to de-escalate medical treatment, specifically chemotherapy.

In the early 2000s, researchers looked at individual tumor

profiles to assess their aggressiveness. The most common type of breast cancer, which makes up approximately 60 to 70 percent of all newly diagnosed tumors, is hormonally fed, estrogen and/or progesterone positive and HER2/neu negative. This type of cancer typically responds to antihormonal therapy, a pill that is relatively easy to take. While all medications have side effects, antihormonal therapy has far fewer than chemotherapy. The question was this: Were there some patients for whom the antihormonal therapy could be enough, or should it be supplemented with chemotherapy indiscriminately? Was there a way to figure out who would benefit from the added, more aggressive therapy and who would likely be cured without it? These questions led to the development of game-changing tools that stratify individual patients into different risk groups for recurrence. There are two main tumor profile tests that give us this information: Oncotype and MammaPrint. The Oncotype test looks at the expression of twenty-one genes in an individual's tumor tissue and generates an aggregate number score that falls into one of three categories: low, intermediate, or high risk for recurrence. The MammaPrint test looks at a seventy-gene profile and stratifies risk groups into either low or high. Importantly, these tests are called *genomic* profiles, run on the actual tumor tissue, and should not be confused with *genetic* testing done to assess for cancer predisposition, the BRCA genes and others discussed in Chapters 11 and 12. Today, many patients with low or intermediate scores can now safely avoid chemotherapy.

The first study validating the utilization of a recurrence score test was published in 2004 and slowly started to affect practice patterns related to recommendations for chemotherapy. Initially, the test results were validated only for patients with negative lymph nodes, so if you were found to have any spread to the lymph nodes, chemotherapy was still given no matter what. The results of a subsequent trial, TAILORx, involving over 10,000 patients with negative nodes across seven countries was published in 2015. This trial demonstrated that a low recurrence score of 10 or less was associated with only a 1 percent chance of recurrence in most cases and, equally as important, no further benefit of chemotherapy. A follow up publication in 2018 also showed no benefit of chemotherapy with high survival rates for most of those in the intermediate risk group with scores of 11 to 25. This was big news and led to a change in practice affecting even more women, decreasing the recommendation and delivery of chemotherapy in this group as well.

A subsequent study called the RxPONDER trial demonstrated that for postmenopausal patients with one to three positive nodes, the test was also valid. In this additional group of patients, chemotherapy could also be avoided if patients were found to be at low risk for recurrence, based on their Oncotype test results. Patients found to be at high risk of recurrence would still need chemotherapy no matter what. It's true that the Oncotype and MammaPrint results sometimes lead to patients with small tumors who would not have been given chemotherapy now receiving it due to a high risk of recurrence. However, the *net* effect of introducing this test into practice more than twenty years ago has been an overall *reduction* in giving chemotherapy by more than 30 percent, again, without compromising long-term outcomes or survival rates. Critically important is to know that in all of these scenarios, patients with low and intermediate oncotype scores still need anti-hormonal therapy. It's only the added chemotherapy that can be avoided. Patients with high oncotype scores are likely be recommended to receive both. This represents groundbreaking progress: dialing back the utilization of chemotherapy while still delivering equal outcomes for millions of women over the past decade. No longer do women receive chemotherapy purely based on their tumor size or lymph nodes, instead, we treat individuals based on testing their own tumor tissue. These results not only guide treatment decisions but also provide important prognostic information and accurately predict risk for recurrence. In most cases, the Oncotype results give us more valuable and individualized information about a patient's expected outcome than standard tumor staging.

Importantly, Oncotype and MammaPrint testing are valid for patients with estrogen-positive tumors even if the tumor is progesterone negative (approximately 10 to 20 percent of all hormonally fed cancers). Those with progesterone-negative tumors may receive higher Oncotype scores as a group, compared to those with both estrogen and progesterone positivity, and therefore may have a higher chance of receiving a recommendation for chemotherapy based on a higher Oncotype score.

Although the Oncotype test has revolutionized treatment for some patients, it is not relevant for everyone with breast cancer. Estrogen- and/or progesterone-positive, HER2-negative breast cancer is the most common tumor profile, but there are other subtypes of breast cancer for which chemotherapy is still highly likely to be part of the plan. HER2-positive breast cancers often still require chemotherapy, as do the more aggressive triple-negative subtype. Oncotype and MammaPrint tests do not work in these two tumor types and should not be sent as they are not designed to measure recurrence risk for these less common subtypes of breast cancer. Nonetheless, these tests have moved the needle for doctors, and, more importantly, for our patients. The treatment paradigm has changed, and the chances of not needing chemotherapy are better than ever before, with no compromise to long-term survival and cure.

WHEN CHEMOTHERAPY IS REQUIRED

There are a variety of chemotherapy regimens and drugs for breast cancer, from the very aggressive Adriamycin—ominously nicknamed the "red devil" (as Jill Martin referred to it above) for its color in the IV bag—to much milder forms. For those who need it, chemotherapy can be lifesaving, reducing the risk of recurrence and improving overall survival.

After their surgery with me, my patients with invasive cancer get an appointment with a medical oncologist to determine additional treatment. We usually schedule it at the same time as their post-op appointment with me. That's when questions about the next steps usually arise: "Doc, do you think I'll need chemotherapy? What kind do you think they will recommend? I know, the oncologist makes this decision but just tell me what you think. I trust you!" This is where I usually tell them to *not* trust me with making decisions about chemotherapy. It's just not what I do. I don't let the oncologists come into my OR to tell me where to make my incision, so I try to stay in my lane. The decision to recommend chemotherapy or not to, and if so, which kind is complex, individualized, and a product of all of the different treatment options we now have. Having more choices is better, but it's also harder, both for the doctor and the patient. The answer is to work closely with your doctors, always look toward the data, and make the best decision for your individual case.

SIDE EFFECTS

The chemotherapy dread is real. Even though we have done so much to dial down the amount and extent of chemotherapy we give, for those who do need it, the side effects can be significant. There are as many ways to navigate the chemotherapy journey as there are patients who go through it. Some choose to continue to work, while some don't. Some exercise, while others scale back. Some socialize more, while others do less. These are all personal decisions based on how a person feels going into chemotherapy, during it, and on the other side. While there is some overlap between the chemotherapy drugs we use for breast cancer and other types of cancer, such as lung, ovarian, or pancreatic, we also often use different drugs. So if each person's experience going through chemotherapy for breast cancer may be different, that's certainly true when comparing the experience of someone who has been through chemotherapy treatment for some other type of cancer.

Similarly, the side effects from chemotherapy are as varied as the number of agents we have to offer: compromised immunity, low cell counts, fatigue, nausea, weight gain, diarrhea and GI complaints, and, of course, hair loss. Because chemotherapy is usually given in weekly or semiweekly cycles, the side effects can be cyclical too. Patients don't report feeling horrible all the time. For many women, the worst days can be somewhat predictable within each cycle. Some of my patients who want to continue working through chemotherapy schedule their chemotherapy for a Thursday or Friday to experience the worst side effects over the weekend, as you will hear from Gina in "The Breast Treatment Advice" at the end of this section. Some patients are pleasantly surprised by how they handle it, and others feel completely wiped out. My advice for patients starting chemotherapy is to pace yourself, be kind to yourself, and, as best as possible, avoid checking out of your life altogether. Sometimes, nudging yourself to get up and do something, no matter how small, can be just what is needed.

COLD CAPPING TO PREVENT HAIR LOSS

Perhaps the most dramatic side effect of chemotherapy is hair loss. It's difficult to hide. It usually involves eyelashes, eyebrows, and everything that has a hair follicle. That's why a device called the "cold cap" was developed. The cold cap goes on a patient's head just before chemotherapy, and the cold effect to the outside of the head slows blood flow to the scalp, reducing the amount of chemotherapy that reaches hair follicles. The net effect increases hair preservation. While there is always some hair thinning, the amount retained usually allows a woman to get through chemo without needing a wig. The cold cap has been widely used for the past twenty years in the United States. There was some concern initially that the utilization of the cold cap might create a "reservoir" in the scalp for tumor cells, where chemo would not reach them, increasing the risk of recurrence. While there have not been any randomized studies, with long-term experience, we can now safely say that the data point toward no increased risk of recurrence or scalp metastasis with cold cap usage.

Here are some caveats for cold capping:

1. It doesn't work with *all* chemotherapy regimens. For many of the stronger treatments' courses, the amount of hair loss is expected to be high, and the cold cap usually won't make a big difference.

2. It costs more. Until recently, cold capping was not covered by any insurance. Thankfully, that has changed, and coverage is now more widely available. I am proud to say that Mount Sinai was the first hospital to offer cold capping to all of our patients, regardless of their ability to pay, thanks to a special

cold capping fund generously established by donors to our hospital for everyone who wanted it.

3. Cold capping can be hard to endure. You have to wear the cold cap for thirty to sixty minutes before treatment, during the entire treatment, and for one or two hours to follow, thus very much prolonging "treatment" time. Many women feel genuinely cold throughout their bodies while wearing the cap. They describe it as the worst ice-cream headache you have ever experienced, lasting for hours at a time.

4. Women can't color their hair or aggressively style or brush it during chemo and cold capping since even the retained hair is very fragile.

5. For women who retain the majority of their hair, the cold cap is a game changer, and I encourage all my patients who do need chemotherapy to consider using it, as you heard with Jill, whom I told to "Sit in the chair, and wear the cap." I know it sounds harsh, but her dread of hair loss, which is common for so many of my patients, made me want to encourage her in a way I knew she would respond to. If you haven't seen it already, I recommend that you watch the video of her on TV right in the middle of treatment, showing all of America the impact of the treatment, and the cold cap, on her hair.

Remember, if you need chemotherapy and the plan is for a regimen that does involve hair loss, make sure you ask about the cold cap if preserving your hair is important to you. After all, not everyone cares about this side effect equally. As with all side effects of chemotherapy, your oncologist will work with you to manage hair loss and other side effects.

25

CHEMOTHERAPY FIRST OR SURGERY FIRST?

PATRICIA K., special needs teacher: Most of the people I know who had breast cancer got surgery first. I needed to start with chemotherapy for my specific type and stage of cancer. That was weird for me. Everyone else I know got a minute to prepare for treatment; they had their surgery first, recovered from that, and then started treatment. Some didn't need chemo at all. I needed it, and my treatment was like "BOOM." I understood why I needed it first: the cancer had already spread to the lymph nodes under my arm, and I was triple negative. Since I was young and healthy, my doctors wanted to hit it hard and right away. From the time I first saw my doctors to the day I started treatment was less than three weeks. I had so little time to prepare for what I knew would come with the treatment: the feeling crappy, the hair loss, the weekly appointments that lasted hours. I was also in a clinical trial, so I got this new drug on top of the regular treatment. I was nervous about it, but the trial had been going on for a while. I knew that Black women are not often included in clinical trials and that there is not enough information on how to treat us, Black women, specifically. Do we have different side effects? Do our tumors respond as well? I wanted to be part of the solution to that lack of information. I also know that the medications I received that are now standard were part of someone else's clinical trial before me, so I wanted to pay it forward. Because I responded so well to the chemotherapy, I got to have a lumpectomy instead of a mastectomy, which I probably would have needed if I had surgery first. I did not have to have all my lymph nodes removed because the spread melted away there too. I'm a teacher for children with special needs, so my job is demanding mentally and physically. Removing all my lymph nodes would have meant a longer recovery and potential for lymphedema, which could have affected me getting back to those kids.

GINA S., media professional: Chemo was such a scary thought for me. With triple-positive breast cancer [estrogen, progesterone and HER2/neu all positive], I knew that I was going to need chemo no matter what because of the HER2/neu-positive part, but it was really scary for me to think about it right away. I believe the standard is if the tumor is two centimeters or more, it's usually chemo first. Under two centimeters is a gray area for my kind of cancer, in terms of whether to start with chemo or surgery. My MRI came back with an estimated tumor size of about 1.5 centimeters. It was very borderline. Dr. Port said, "I think we can get this out with surgery first," but it was ultimately my choice. That was enough for me to say, "All right. Let's try it." I loved the fact that I was given a choice, but I also know that every patient is not the same. I knew going in that the whole thing might not come out and that I might have to readjust. Thankfully, I got clean margins, and I was very happy with my decision to do my lumpectomy surgery first. Having the surgery done, knowing the tumor was out and I had clean margins, meant I could think clearly again. That gave me time to prepare for chemo.

STACEY GRIFFITH, founding senior master instructor of SoulCycle: I did a clinical trial with chemotherapy first, and it turns out, I was cancer free before the surgery. The chemotherapy killed all my cancer and allowed me to have less aggressive surgery. When I did my lumpectomy and we checked my nodes, there was no cancer left. I love the concept that doing less allows you to think about it less. It's not like a new normal that you wake up to. It's your normal.

Now it doesn't even cross my mind on a daily basis that I ever had that experience. That thirteen months of my life is not even in a file in my brain anymore. It's not even in there. It's just gone. I've moved on and I feel amazing. I'm doing a hundred push-ups a day.

DR. PORT: If you are newly diagnosed with breast cancer and choose to share your story, you will hear the stories of so many others who are going through or have been through breast cancer. As you start paying attention to different journeys through surgery, radiation, chemotherapy, antihormonal therapy, and even newer treatments like immunotherapy, chances are that most women will have had surgery first, followed by whatever treatments they needed afterward. That's because surgery accomplishes so much: it can get the vast majority, if not all, the cancer out. For some women, it's even curative by itself. A surgery-first approach can provide patients and doctors with information to guide subsequent treatment. In the prior chapters, we discussed de-escalating treatment for those who may not need it. Those kinds of decisions can really only be made knowing the true nature of disease, how much is there, and whether or not it has started to spread. In most situations, the best way to figure that out is through surgery. For example, in the last chapter we discussed how the Oncotype test can have significant implications regarding whether a patient needs chemotherapy. But that test is best performed on tissue obtained through surgery. Plus, knowing if the lymph nodes are involved can help us decide whether we should even be doing an Oncotype test at all. To put it simply, what we learn in surgery informs downstream treatment decision-making.

However, an increasing number of women—now 20 to 30 percent—are actually getting medical treatment first before surgery. In many cases, starting with treatment also provides valuable (but different) information. Think about it: if you remove the cancer in surgery and then give treatment to wipe out any cancerous cells, how can you really know if the treatment is working?

The good news is that for hormone-positive cancers, we know the antihormonal treatment we give is very effective in the majority of cases. That treatment goes on for years, so starting with it when a patient needs surgery, too, would be impractical. For other subtypes of cancer, like HER2 positive and triple negative, assessing response to treatment by giving it first can provide invaluable information. In certain types of breast cancer, giving treatment first can contain and control disease that might have spread or is at a high risk for spreading. Here's some terminology you may hear your doctors use: chemotherapy after surgery is called adjuvant treatment. Chemotherapy before surgery is called *neo*adjuvant therapy.

Let's go subtype by subtype and run through the general guidelines for when treatment first may be a better approach for each type of breast cancer.

TRIPLE NEGATIVE

A triple-negative breast cancer diagnosis is one of the most common reasons to consider chemotherapy first. We recommend neoadjuvant treatment in patients with tumors that are two centimeters or greater, and/or have lymph nodes already involved, which we can learn from a needle biopsy of the lymph nodes under the arm. For patients with tumors that are thought to be less than one centimeter and have no obvious nodal involvement, we usually do surgery first. For patients between one and two centimeters, we determine whether to start with surgery or chemotherapy on a case-by-case basis.

There are four benefits of starting with chemotherapy in this subtype:

1. Assessing response to treatment by monitoring along the way. When patients are having their treatment, we can do imaging studies and exams during treatment to actually see and feel the cancer shrinking. We can also determine if there

is *any* residual disease left when we do our surgery to follow. If so, we still have the option to give more treatment after surgery.

2. We can contain and control disease that is more aggressive and has a higher likelihood of spreading. Taking a tumor out of the breast does not affect cells that might have already disseminated (which we can't test for), but chemotherapy does.

3. If a tumor shrinks down and responds to chemotherapy, we can switch someone with a larger tumor from a mastectomy to a lumpectomy, allowing them to have less extensive surgery when that time comes.

4. We can convert someone with positive nodes to negative nodes, also called downstaging—much better than upgrading—reducing the scope of surgery under the arm and the risk of lymphedema. If a needle biopsy finds positive nodes when a person is diagnosed, we can check that specific node for residual disease at the time of surgery. If that lymph node looks clear under the microscope, implying that the treatment eradicated all cancer in the lymph nodes, we don't have to take out the rest. If the nodes are still positive, we usually still remove them.

In approximately 50 percent of patients with triple-negative breast cancer, neoadjuvant chemotherapy will result in what's called a CR, or complete response. That means that when we go to surgery to remove the area and nodes, there is no residual cancer left. For those with remaining cancer, we have the option to give more treatment to follow, knowing some or part of the cancer was resistant to the treatment that was already given.

HER2/NEU POSITIVE

The approach to HER2/neu-positive disease is similar to that of triple negative, and the criteria for treating with medicine before surgery are basically the same. Since we have targeted therapies for HER2/neu-positive breast cancer, the complete response rate is higher than triple negative—approximately 60 to 70 percent—as we saw with Stacey's story.

ESTROGEN/PROGESTERONE POSITIVE, HER2/NEU NEGATIVE

For this most common subtype of cancer, we usually take the surgery-first approach. That's because these tumors don't usually respond as exuberantly to up-front chemotherapy; the complete response rate is only about 20 percent. That means that there is very little chance of shrinking the tumor size to facilitate a lumpectomy or downstaging lymph nodes to reduce the need for axillary dissection—removing all the lymph nodes. Because chemotherapy is given in cycles and usually lasts months, we don't want to be giving months of treatment that doesn't have a high likelihood of being effective. People who *do* need neo-adjuvant chemotherapy in this group are mainly women who have cancers that are what we call unresectable, meaning that we couldn't remove it all in surgery, even if we tried. A cancer might be unresectable, even with a mastectomy, if it involves the overlying skin or invades the underlying chest muscle. Extensive lymph node involvement with nodes that are matted together and cannot be fully removed would make the cancer unresectable. Cancers that have spread to other lymph node areas, like the internal mammary nodes in the chest, which we can see on a

PET scan, are not removable with surgery. Any situation where I'm worried that I might not be able to remove the entirety of the disease would lead me to recommend chemotherapy first and divert the patient away from having surgery up front. Surgery to remove just part of the cancer is not beneficial to anyone. Over the past twenty-five years, I've only had two or three occasions where I've gone into surgery and realized I couldn't get it all out. That only happens when all of the imaging, scans, and exams that we do prior to surgery underestimate the true extent of the disease. As a surgeon, I can tell you that this is one of the most demoralizing situations to be in. When it happens, we usually leave markers in the region where the disease could not be removed and hope that additional treatment like radiation can eradicate the rest.

Ultimately, most patients with breast cancer do have surgery first, since the majority have the type of cancer where surgery first is the best approach. But for women with other subtypes—HER2 positive or triple negative—the best approach has to be decided on a case by case basis, understanding the overall benefits of each pathway. Surgery first provides information that can help impact plans for further treatment, if there are decisions to be made. Treatment first can show us the response to therapy that, in turn, can dictate whether a person needs even further therapy to follow. It can also shrink disease to possibly allow for less extensive surgery. This decision-making process truly exemplifies our movement past a one-size-fits-all approach and toward precision care. For patients with breast cancer, the age of personalized medicine is now.

26

RADIATION

HANNAH STORM, sports anchor, producer, and director: My diagnosis was a total shock when it came at age sixty-two, two decades into getting regular mammograms and never having anything suspicious come up. I do have dense breasts, so I follow my mammogram with an ultrasound every year, which is when we realized there was something that we needed to follow up on. By law now, everyone has to be informed if they have dense breasts. What you aren't always informed about is that *you* need to pick up the phone and schedule an ultrasound.

After the ultrasound, I had a biopsy, which came back with DCIS. I had zero history of breast cancer in my family and very little history of cancer of any kind in my family. Thus, after so many uneventful mammograms, it came as a surprise.

The good news is, with early detection, DCIS has a 99 percent survival rate within five years. When I called Dr. Port, she said, "You're not going to die from this." That made all the difference in the world. Immediately, being a sports person, of course I was like "What's our game plan?"

We made a series of logical steps to determine what we were going to do moving forward. One of those steps was genetic testing, something I knew nothing about at the time, but I learned it was really important to determine the type of surgery I would have and how aggressive we would be. Was I genetically predisposed to getting cancer again, for instance?

Because my genetic profile was such that I didn't have any red flags, we were able to look at a lumpectomy. The other really critical step that we took was to do an MRI to see what we were dealing with. It turned out we were not just dealing with one tiny area, but two. Luckily, these two areas were adjacent, and we were able to determine that these two could be grabbed in one fell swoop.

After surgery, it was explained to me that we weren't going to jump into anything. We were going to wait and see what kind of story my cells were telling us. So, a test was done to determine the aggressiveness of my DCIS and the relative benefit of radiation to follow, which is often part of the treatment course after lumpectomy. This is the type of testing that I would not have known to ask for that the radiation doctor recommended and had done. Sure enough, I was not a candidate for radiation. The test results were easy to understand. It was a number scale. It showed that had I undergone radiation, it would not have benefited me. Getting the right kind of testing helps ensure that you're not going to be over-radiated or overmedicated, where you're living a menopausal life all over again. If possible, you're going to come out of your surgery and be able to move on with your life. That was the case with me. I do take "baby tam," a small dose of tamoxifen. I have a few side effects but they're absolutely manageable.

I'm also really grateful that I have my yearly mammograms. After twenty-two mammograms and ultrasounds, it doesn't mean you're free and clear for the future. You can get cancer at any time. It does not matter what your family history is. Anyone can get cancer at any time. It does not discriminate.

DR. PORT: The change we've seen in radiation treatment is emblematic of the way we've moved past a one-size-fits-all approach. We went from "Radiation for everyone" to "Let's dial down the length and extent of treatment and see if that works just as well" to "Hey, maybe not everyone needs it, so can we figure out through research who those people might be?"

The kind of radiation we use for breast cancer is usually what we call external beam radiation, given from a radiation machine and pointed at the breast. Radiation involves a series of treatments where a person lies on a table, either face up or prone, while a beam emits radiation focused on a specific organ, like the breast, or a specific spot within that organ. Typically, we give radiation treatment to the whole breast, plus a boost directly to the lumpectomy site. Radiation tends to happen in short sessions, approximately ten minutes each, once a day, five days a week for a given number of weeks that can vary. The goal is to eliminate any microscopic cells left behind in the remaining breast after a lumpectomy, the remaining skin and chest wall after a mastectomy, or the remaining lymph nodes under the arm and in the chest.

Patients often mix up the side effects of different cancer

treatments. Every day, I hear patients saying things like "I don't want radiation because I don't want to lose my hair!" Hair loss can be a side effect of chemotherapy, but it's not at all a risk with radiation. You don't lose your appetite or experience any significant effect on any other part of your body. People who've gone through radiation with other types of cancer might tell you to beware of side effects like GI distress or inability to swallow. This is when it's so important to tune out the noise of misinformation. Radiation for thyroid cancer may involve quarantining away from other people, especially small children, since patients are, in fact, radioactive. Radiation for esophageal cancer *can* make it difficult to swallow, since it's targeted at the neck. Radiation to the pelvis or abdomen for uterine or rectal cancer can cause lower abdominal and bowel symptoms. None of these have anything to do with radiation as a treatment for breast cancer.

Here are the potential side effects for breast cancer radiation:

1. Patients may experience some minor tanning and burning of the skin, which is temporary and goes away with time.
2. People can get tired toward the end of radiation treatment. Radiation does cause cell damage, since we're targeting bad cells. Just like after surgery, your body senses this damage and goes into repair mode. That bodily repair involves energy, which sometimes means less energy for daily activities. I tell my patients, particularly older ones, not to be surprised if, toward the end of the day during radiation, especially toward the end of the course of treatment, they feel tired. It seems strange that keeping all daily activities the same but tacking on a ten-minute session of radiation would make you

feel exhausted, but it happens, and it's a normal part of the process. I tell my patients to give themselves a break, listen to their bodies, and know that the energy drain will go away.

3. While radiation cures cancer, it can also cause cancer. There are risks of developing what's called a "secondary malignancy," cancer related to previous cancer treatment, from radiation. These risks are small—less than 1 percent—but serious. Radiation can cause rare tumors of the chest wall at the site of treatment. The main type of cancer we see after radiation therapy is called angiosarcoma, tumors that can be very aggressive and difficult to treat. Again, and I cannot overstate this, these tumors are extremely rare with a risk of less than 0.05 percent, but they're important to know about going into radiation.

4. Damage to the heart and lungs is often brought up as a concern for giving radiation, especially for patients who may already have underlying heart or lung disease. It's true that the heart and lungs are directly behind the breasts. Does that make women who have left-sided breast cancer more at risk for heart effects, since the heart is on the left side of the body? No, it doesn't. With focused radiation techniques, our radiation oncologists and the technologists who help plan patients' treatments can steer away from the heart and lungs to focus on the breast. In keeping with no the one-size-fits-all approach, of course, there are some patients with underlying diseases for whom even a small exposure to radiation may be damaging. In that case, I'd worry more about that disease and less about the damage of radiation exposure. Our radiation oncologists, the doctors designing and giving the radiation treatment, can always consult with their colleagues, other specialists, about risks and benefits to a particular patient

based on her underlying conditions before giving the treatment.

5. Finally, radiation can make a future reconstruction of the breast more challenging. When a woman has a lumpectomy and radiation, her chances of the cancer coming back are quite low, but low does not mean zero. If a person has a recurrence of breast cancer after a lumpectomy *and* radiation, the standard approach, in most cases, is to do a mastectomy. While radiation of the skin has a low side effect profile, it can affect the skin on a microscopic level, making it less elastic and damaging small vessels that affect wound healing. These factors can compromise the success of reconstruction. Thankfully this scenario is rare, and usually no further surgery is needed on the breast after lumpectomy and radiation.

Ultimately, radiation is quite safe. It's a critical weapon for treating breast cancer and preventing recurrence. In the age of de-escalation, we don't want to give too much radiation—or any, for that matter—if we don't have to. In fact, as in Hannah's case, there has been significant progress in de-escalation, for both invasive cancer and DCIS, on the radiation front in the last decade.

SHORTER COURSES

Until about fifteen years ago, the standard radiation treatment was five days a week for six weeks. That course of radiation was given to almost all breast cancer patients who had lumpectomy, as well as patients who had mastectomy and needed radiation to follow. In the last decade, however, clinical trials have shown that shorter courses, called hypofractionation, are just as safe. Hypofractionated radiation essentially gives the same dose *overall* but divides the dose among a smaller number of treatments, giving a larger dose at each treatment, and shortening the overall course. This allows a woman to complete her radiation in three to four weeks, rather than six. Those two to three weeks may sound trivial in the overall scheme of things, but consider this: where I live in New York, there is a radiation facility within a stone's throw of almost anywhere a person might live or work. That's not true for more rural parts of the country, where driving to the closest radiation facility could take hours. Hours of driving for a ten-minute treatment every day of the week sounds crazy, but you may need it if you want to save your breast and have a low risk of the cancer coming back. For a woman with a long commute to radiation, shortening the course can be the game changer that lets her complete her course of treatment expeditiously and get on with her life, her job, and her family. Most importantly, this shortened course of radiation has been shown to be effective across most age groups, most tumor types, and most subtypes of breast cancer. Some women with more extensive or aggressive disease do still need the full six weeks, but so many now don't, which is a huge step forward.

SHORTER STILL . . .

We also have an even shorter course of radiation called accelerated partial breast irradiation (APBI), which can be completed in a week or even less. For some patients the lengthier, stronger treatment—even if it's only three weeks—can be overkill or difficult to get to. So APBI is even shorter still. Some forms of

APBI can actually be given *directly* into the surgical site. With intraoperative radiation (IORT) a device can be placed in the lumpectomy cavity during surgery to deliver a course of radiation treatment to the tissue directly surrounding the area of the cancer that was removed. If a patient has difficulty traveling, she can come out of her operation having had both surgery and radiation.

Unlike standard radiation, APBI *only* radiates the area of the lumpectomy and the immediately surrounding tissue, rather than the whole breast. When breast cancer recurs, it usually develops in the immediate vicinity of where the cancer was. So, for small, nonaggressive breast cancers with low propensity for spread into the surrounding tissue and no lymph node involvement, radiation to the specific cancer area alone may be enough.

NO RADIATION AT ALL

The ultimate in de-escalation is no treatment at all. There has been significant research and clinical trials devoted to teasing out which patients can safely avoid radiation altogether.

For invasive cancers, it's mostly women over the age of sixty-five who can avoid radiation therapy altogether, since the trials showing acceptably low recurrence rates without radiation have primarily involved women over the age of sixty-five or seventy. Other important criteria that might make you a good candidate for omitting radiation are small tumors; node negativity; and hormone-positive, HER2-negative cancer, which means you can take antiestrogen treatment, which will lower risk of recurrence. Clinical trials show that for women who meet these criteria, the risk of recurrence was approximately 10 percent with omission of radiation, compared to a risk of less than 5 percent with it. By the way, those criteria encompass a significant segment of women who get breast cancer. The disease is most common between sixty and seventy years old, 60 to 70 percent of tumors have this less aggressive profile, and with early detection, tumors are usually small and node negative—yet another example of how screening and early detection allow us to de-escalate treatment on all fronts.

The potential for de-escalation and omission of radiation therapy is also gaining momentum for patients with DCIS. Hannah's story is a perfect example of how we now have reliable tools for predicting the benefit of radiation or lack thereof based on an actual sample of the patient's DCIS. Radiation for DCIS is a potentially important component of treatment that reduces the risk of recurrence, and, importantly, *invasive* recurrence, but it isn't always necessary. With DCIS, the four factors we consider that can influence risk of recurrence both with and without radiation are the following:

1. Patient age: The older you are when you're diagnosed, the lower the likelihood of recurrence over the rest of your life.

2. DCIS grade: DCIS can be low, intermediate, or high grade. Low- and intermediate-grade DCIS may be more amenable to not having radiation, compared to high grade.

3. Amount of DCIS present: After a biopsy shows DCIS and the person has surgery, we can assess the total amount of disease. In some situations, the area of DCIS may be so small that the biopsy removed it all. In these cases, additional radiation to mop up residual disease in the surrounding tissue may be superfluous.

4. **Margins:** While age of the patient and the grade of DCIS, as seen on the biopsy, can usually be determined before surgery, the actual amount and the margins can only be determined through surgery. The pathology report from surgery will provide us with this information by looking closely at all of the edges to make sure they are clear, ideally by at least a two-millimeter margin. Adequate margins do translate to a lower risk of recurrence and are critical for patients considering omission of radiation. If we give radiation to mop up the microscopic disease left behind, having a larger amount of DCIS or close margins are predictors of residual disease.

If you are a patient with DCIS who clears all four of those hurdles—you're over the age of sixty with a small amount of low- or intermediate-grade DCIS and wide margins—the decision to not pursue radiation is very reasonable. Most patients don't meet all four of the criteria. Fortunately, there are newer, more personalized testing options that go beyond the checklist and help us make these decisions with our patients. We can send a piece of the tissue removed during the surgery to one of a few different companies that provide information on the risk of recurrence and the benefit, if any, of added radiation. With more information, we can practice more effective decision-making. A patient might choose to opt out of radiation based on this information, but she might still want to pursue it. With a high risk of recurrence but no added benefit of radiation, she might even revisit the option of mastectomy. The development of these tests through research represents significant progress over the last ten years. They are added value and information toward more precise decision-making regarding optimal treatment for DCIS on an individual basis.

YOU CAN ONLY HAVE IT
ONCE. OR CAN YOU?

The standard of care treatment for someone who has had a recurrence after lumpectomy and radiation has traditionally been a mastectomy. That's in part because the rule of thumb had been that we can't give radiation to the same breast twice; the skin damage and other toxicities would be too great. Now that we have newer regimens for radiation that do not involve full-blast treatment, treating recurrence with mastectomy is no longer a foregone conclusion. There are more options than ever before for a woman who has had radiation and develops a recurrence or new cancer on the same side. The conversation here is a nuanced one and should involve a multi-disciplinary discussion and approach with input from lots of specialists. Can the radiation doctor safely reradiate, especially if it was a decade or more after the first when much of the skin and tissue effects may have subsided? Might it be safe to do no radiation at all for this new cancer? Are there medicine options that would reduce the risk of recurrence, if no radiation was administered? All of these factors must be considered when deciding about treatment for recurrence. These decisions have made our specialties in breast surgery, oncology, and radiation more nuanced than ever before, and drive home the benefit of seeing true experts who have experience with these complex decisions and can work together—with you—to come up with the best plan.

WHEN YOU STILL DEFINITELY NEED IT

In general, for women with more aggressive subtypes of cancer, like HER2-positive and triple-negative cancer, radiation is necessary after a lumpectomy. The risk of recurrence is higher, and forgoing radiation increases that risk significantly.

Lymph node positivity will also often warrant additional radiation. Women with four or more positive nodes will get radiation, regardless of mastectomy or lumpectomy, and women with one to three positive nodes will often be recommended to receive it as well, depending on other high-risk features. Newer trials have shown that radiation treatment is just as effective as removing all nodes under the arm, or axillary dissection, for women with one to three positive nodes—and it's associated with a lower risk of lymphedema.

Women with advanced disease who have a mastectomy with large tumors or multiple lymph nodes involved may often require radiation, even after mastectomy. This is referred to as post-mastectomy radiation treatment or PMRT. Yes, even with excellent screening options, which translate into the best chances for early detection, there are still many cases of advanced disease that require full courses of treatment on all fronts.

WHEN YOU CAN'T HAVE IT

There are certain situations where radiation, either as a repeated treatment or not, is simply not advisable. Breast conservation therapy usually involves the combination of lumpectomy *and* radiation. If you can't have the radiation, and it would be necessary in your case, you will probably be told to consider mastectomy instead. Thankfully, these conditions are few and far between, but they're still important to know. The first situation is a prior history of chest wall radiation, like having undergone the treatment for Hodgkin's lymphoma. Young women who develop Hodgkin's disease and need chest wall radiation (also known as mantle radiation) are at a significantly higher risk of getting breast cancer in the future, starting at about ten years after treatment. Developing breast cancer *after* having received chest wall radiation makes it difficult to radiate again. Plus, the radiation exposure usually means a higher risk of getting yet another breast cancer in the future. All of these factors can make a second round of radiation more challenging, and that's why many women in this scenario are recommended to undergo a mastectomy or even a bilateral mastectomy.

The second main group of patients for whom radiation may be too risky are those with serious connective tissue diseases like lupus and scleroderma. These diseases have their own effects on the skin and soft tissue that can be further exacerbated by radiation and can have severe adverse effects on the underlying disease. Recently, I saw a patient with severe scleroderma who was told in no uncertain terms by her rheumatologist and the radiation oncologist that radiation would not be advisable. Despite having a small tumor that would have been an easy lumpectomy from a surgical standpoint, we recommended a mastectomy.

Finally, even though radiation focuses on the breast and soft tissue and does not reach the underlying heart and lungs in appreciable amounts, there are patients with such significant heart and lung disease that even a small amount of exposure might be ill-advised.

As I counsel my patients on which type of surgery to have, I'm always factoring in information about the whole person, including any underlying conditions she may have. Patients are not just their breasts. If we cure breast cancer but exacerbate some other life-threatening condition, we have not done our jobs. With every patient, we consult their other doctors to add to our knowledge about a patient's overall condition and well-being. So frequently, there are non-breast-cancer-related health issues that tip the scales on decisions about surgery and downstream treatment.

I don't describe all of these scenarios to diagnose you or advise you on your particular case. I do it to give you some general ideas for points to discuss with your own doctor. Some of the scenarios I've outlined above may resonate with you, but it's important to know that there are other things about you as an individual that may make your case different from someone else's, even someone who seems to have "the exact same thing." There are also other data points that doctors factor into their recommendations. Those might include Ki-67, a measure of tumor aggressiveness, and lymphovascular invasion, a measure of propensity to spread. The bottom line is that no one's case is just one factor. The specialist's job is to pull all of the factors into a composite picture and make recommendations that give you the best chance for the best outcome. Now more than ever, we have a wide array of data points and better tools to help make these decisions together.

27

ANTIHORMONAL TREATMENT

MARY JOE FERNÁNDEZ, ESPN television commentator, tennis Olympic Gold Medalist, three-time Grand Slam singles finalist: In the beginning you're prepared for everything. Am I going to need radiation? Chemo? When I learned we have to wait for the Oncotype number to come out to make a decision, I prayed for a low number. It was such a blessing for me not to have to do chemo. I had to take tamoxifen and I read all the side effects with Dr. Port. "It says 'causes death,'" I said to her, and she responded, "Yeah, so can crossing the street." It's true, you could get hit by a car too. You have to weigh the pros and cons of whatever your treatment plan is. You make the best decisions with the information you have.

My advice is everybody is different, everybody reacts differently to the medicine. I was totally fine on it. For me, it was a mentality thing: *It's doing good for me, it's keeping my cancer away.* I luckily didn't have any side effects so I accepted it and was on it for over seven years.

MOLLY W., social worker: When I was first diagnosed with breast cancer, I had no idea what it would mean for me. It's not like I had other family members who had gone through it or who could tell me what to expect. Even if I did, their stories, as I learned, would not necessarily be mine.

As I found out more about my cancer, it became clear that the road ahead could involve multiple steps of treatment and that it wouldn't be something I could just power through over a few days or a week. It was so overwhelming at first. When would I have surgery? Would I need radiation? Would I need chemotherapy? I was dreading chemotherapy beyond anything. Most of all, there was the waiting. It seemed like every time I cleared one hurdle, there was yet another test or piece of information I had to wait to hear about—and stress over the outcome in the meantime.

In the end, because of newer research and tests, I didn't need chemotherapy. I couldn't believe it! I did have some cancer spread to my lymph nodes, and I had thought that even a small amount of spread would automatically mean chemo, but that turned out not to be true. Look, I wasn't going to refuse chemo if I was told I needed it, but, in the end, I was thrilled that I didn't even need the treatment I'd made myself crazy over. Looking back, I wasted so much mental energy and too many sleepless nights worrying about how I would get through. My advice is don't dread what might not even happen. Take it one step at a time and try not to worry about treatments that you may not even need.

Not needing chemo didn't mean I didn't need *any-*

thing. What I did need was a hormone blocker. My cancer cells fed on estrogen. Arimidex, a drug I would take for five years, would block estrogen production in my body. If any cancer cells remained or escaped, they wouldn't get any of the estrogen they need to live on. In other words, the escapees would die. Arimidex has side effects, but I find them pretty tolerable. I am achy in the mornings, until I get going, and I get an occasional hot flash. I think of it as my lifeline: a friend who's got my back. Taking the medication is a small price to pay, in my opinion, for keeping my cancer from coming back—and for not needing to do chemo. It's a win-win for me.

AMANDA KATZ, certified clinical medical assistant: I saw the oncologist and she gave me my Oncotype. It was six, which was low. That was awesome to hear because I didn't need the more extreme treatments like chemo or radiation. *I'm really getting off lucky,* I thought. It wasn't like, *Why me, why did I get cancer?* It was like, *Why me, why did I get so lucky?* Why did I get chosen to skate through almost unscathed, except for the physical change and now being placed on tamoxifen, which for me, was a killer of a medication, at first. It really messed with my brain.

I didn't have any night sweats or any of that stuff before. I had a regular period. All of a sudden that changed. For the first couple of months, it was dramatic: night sweats, hot flashes, emotional ups and downs. I flipped out on my boss once because he didn't thank me for doing something. I was more upset about it than I was about getting cancer, weird. It was so unlike me. It was a trippy time. It took about three months to normalize with my emotions. Now I'm okay with it.

DR. PORT: Antihormonal therapy has been around for about fifty years. It comes in various forms, but the majority of them are medicines that reduce the risk of breast cancer recurrence by 50 percent or more for patients with hormone receptor–positive breast cancer. As doctors, we can't know if cells have already spread even before surgery, so these medications are our insurance in case they do. For women with an extremely low risk of spread, the addition of these medications may be overkill, but the benefits are significant for most women with hormone-positive invasive breast cancer. The medications have three potential benefits:

1. Reducing the risk of recurrence in the breast in women who have had a lumpectomy.
2. Reducing the risk of developing a *new* cancer in the other breast in women who don't have a double mastectomy.
3. Reducing the risk of developing metastatic disease and therefore the risk of dying from breast cancer.

Since 60 to 70 percent of breast cancers are driven and fueled by hormones, hormone-blocker medications, also called antihormonal therapy or endocrine therapy, can be a critical part of treatment for the women who make up that statistic. These medicines fight cancer by depriving them of the estrogen supply, which the majority of breast cancers need to grow and live. I explain the hormone blockade to my patients using a plant-water analogy. Imagine that your cancer is a plant—or, more fittingly, an unwanted weed. That weed, like all plants, needs water to live. For breast cancer cells, that water is estrogen: the hormones circulating in your body, coming from your ovaries until menopause. Even though estrogen levels drop off after that, our postmenopausal bodies still make estrogen in the adrenal glands, fat stores, and even the skin, in small amounts. Estrogen, regardless of where or when it's produced, can fuel breast cancer cells that are what we call hormone positive.

In most cases of invasive cancer, the medications are taken for five years, although the exact timelines can vary. The benefits of the hormone blockers extend beyond those five years. We may recommend that women at higher risk for recurrence stay on the medication for up to ten years. The decisions of which medication to take and for how long should be made in consultation with a medical oncologist. My advice is to seek out a medical oncologist specialized in and focused on breast cancer. There are new studies and options almost every year, and it's hard to stay up-to-date on every kind of cancer, as a general oncologist would have to. A specialized breast medical oncologist can give you the best advice.

SERMS: TAMOXIFEN AND OTHERS

The history of hormone-blocking medication begins with the observation over a century ago that removing the ovaries often led to regression of breast tumors. Based on this initial observation, in the 1960s and '70s, a medication called tamoxifen was developed. This medication falls into a category called a selective estrogen receptor modulator (SERM) and binds directly to the estrogen receptor in tissues all over the body, including the breasts. Interestingly, that binding process has different effects in different types of tissue. While it serves as a hormone blocker in breast tissue, it's actually a hormone stimulator in other tissues—such as uterine—ergo the name *selective* estrogen receptor modulator.

Tamoxifen has probably prevented more breast cancer deaths than any other single medication, due to its effectiveness and the longevity with which it has been prescribed. Looking back at the last fifty years, the total number of lives saved by this medication *alone* are incalculable. When we prescribe tamoxifen after surgery for invasive breast cancer, it reduces the risk of recurrence and breast cancer death by approximately 50 percent. Let me make this perfectly clear: if, based on your specific case, your risk of dying from breast cancer is 20 percent, tamoxifen reduces that risk to 10 percent. If your likelihood of survival is 80 percent, tamoxifen increases it to 90 percent. Tamoxifen, and the countless lives we saw it save, was one of our earliest reasons for optimism in the breast cancer field.

While tamoxifen is one of our oldest versions of endocrine therapy, it is also still one of our most utilized and is still the main medication prescribed to premenopausal women with hormone-positive breast cancer.

Like all medications, tamoxifen does have side effects. The serious ones are thankfully rare: a very small increased risk of blood clots; a worsening of cataracts, if you already have them; and an increased risk of uterine cancer, almost exclusively in women over the age of sixty. Then there are the less medically serious side effects that can affect quality of life. Even though tamoxifen is the most lifesaving medication for breast cancer, it also generates one of the highest levels of patient reluctance and concern about side effects of any other medication we use.

The less medically serious side effects of tamoxifen include menopausal-type symptoms, such as hot flashes, night sweats, and weight gain. Some women report vaguer symptoms such as nausea, headaches, brain fog, or loss of mental acuity. Unfortunately, almost any possible symptom that someone experiences while on tamoxifen tends to get attributed to it. To be sure, there are side effects that can impact quality of life, but for most women, they're mild and tolerable. However, if you go to any breast cancer–related chatroom, that does not appear to be the case. As with everything, many women offer helpful coping mechanisms for the common side effects, and what worked for one of them might work for others. But hearing about these side effects of a potentially lifesaving medication is often unhelpful. They only prime the patient to think she will get the horrible side effects she's read about.

I always recommend that my patients minimize the noise about the adverse effects of tamoxifen and try it with an open mind. Because while some serious side effects like getting a blood clot are not influenced by attitude or perception, others, like brain fog and headaches, can be. Attribution of side effects can influence compliance with taking the medication. Oncologists can help women adjust to the medication by starting slowly and ramping up dosage or waiting so other prior treatments and their side effects (like radiation and its potential for causing fatigue) have the chance to wash out. These are examples of strategies that oncologists use to help women adjust to this medication, which, again, has an overall mild and tolerable profile. I always encourage my patients to go into the long haul of tamoxifen with a positive attitude. "If it doesn't work for you, you can stop," I always say. We are cancer doctors; we are not the cancer police. No one comes to your house and checks if you are picking up your prescription or taking your pill. It is important to know that not taking it or stopping tamoxifen means you will lose the improvement in survival rates. That applies to all of the other endocrine therapies discussed below as well.

Because many young women with breast cancer are on tamoxifen, it's important to know that while this medication can cause menopausal symptoms, it does not put a young woman into menopause. She is still potentially very fertile, depending on her age. In fact, it is critically important to know that a woman of childbearing age on tamoxifen can actually get pregnant but should not, as tamoxifen can cause birth defects. So, for women in this age group on tamoxifen, some form of nonhormonal birth control should be practiced, since pregnancy on tamoxifen is a hard no. More on this later in Chapter 32 addressing fertility after a breast cancer diagnosis.

SERDS: FASLODEX AND OTHERS

If a SERM is a selective estrogen receptor modulator, a SERD is a selective estrogen receptor *degrader*. These drugs are similarly used exclusively in patients with hormone-positive cancer, and bind to the hormone receptor, but they degrade it rather than block it. We often use these medications as a second line of defense to tamoxifen's first when women who are already on it have a recurrence—and we're often dealing with more advanced or metastatic disease. The most common SERD is called Faslodex and is given by injection, but oral SERDs can be used and are in further development.

AROMATASE INHIBITORS

Aromatase inhibitors are newer than SERMs and SERDs, but they've been in use for more than thirty years. Aromatase is a critical enzyme in the pathway toward synthesizing estrogen. While SERMs block the estrogen receptor without affecting levels of circulating estrogen, an aromatase inhibitor blocks the actual making of estrogen to shut down the supply. In general, aromatase inhibitors (or AIs, as we call them), are used in post-menopausal women only. They can't really block the amount of estrogen produced by the ovaries in premenopausal women—tamoxifen is still the drug of choice for them—but when ovarian estrogen production has mostly shut down and the main source of estrogen is fat stores, AIs are extremely effective. The main AIs that we use are called anastrozole (Arimidex), letrozole (Femara), and exemestane (Aromasin). The side effects are similar across the board. Menopausal symptoms, even in women who are years past menopause, may return: hot flashes, night sweats, and vaginal dryness. Joint and bone aches, pains, and stiffness may also affect some women, as well as an exacerbation of osteoporosis. That's why women with underlying osteoporosis may need to supplement their AIs with bone-building medications. It should go without saying that every woman experiences these medications and their side effects differently. And even though the medications are supposed to be the same, some individuals have different experiences on each of these different medicines. A woman who has a bad experience on one of them could start taking another and find it better.

While tamoxifen is the medication of choice for hormone-positive breast cancer in premenopausal women and AIs are in postmenopausal women, there are exceptions to the rule. Postmenopausal women who do not do well on AIs and have exhausted all the options can take tamoxifen, although their benefit in improved survival would be slightly less. Similarly, women who are premenopausal with high-risk breast cancer may be recommended to take an AI along with an ovarian suppression medication, called Zoladex or Lupron, instead of tamoxifen. Ovarian suppression medications effectively put a young woman into temporary menopause, thereby making her eligible for an AI. Ovarian suppression medications usually consist of an injection every month, which some women tire of and decide to surgically remove their ovaries; once that makes them permanently menopausal, they no longer need the shots. It's important to know that this ovarian removal is not recommended related to ovarian cancer risk, although it is preventive. Women with breast cancer are not typically recommended to remove their ovaries and don't have an increased risk of ovarian cancer *unless* they are BRCA positive. Outside of being BRCA positive, there is no increased risk of ovarian cancer tied to a breast cancer diagnosis itself, as previously discussed in Chapter 11. There is a lot of misinformation about ovarian cancer risk after breast cancer, perpetuated in part by the fact that some women remove their ovaries to render themselves menopausal, and not related to ovarian cancer risk. Keep this in mind as you hear different stories.

ALL THE LATEST NOVEL THERAPIES

Stacey Sager is a widely renowned, award-winning New York Channel 7 news reporter who has been sharing her story of a three-decade journey with cancer to help empower people. Stacey was first diagnosed with breast cancer at only thirty years old, when she was found to be BRCA positive. She encouraged patients to have genetic testing years before the gene was widely known and she unquestionably saved lives, including those of some patients who came to see me after testing positive for the gene they first heard about from Stacey, before I even knew her. She later had ovarian cancer and, most recently, a breast cancer recurrence at fifty-six, when I became involved with her care. Here's Stacey's account of seeing cancer treatment evolve and being the recipient of a novel medication.

STACEY SAGER, television reporter: I've done the experiential thing three times now, starting in 1999, when I had my first breast cancer diagnosis. At every turn, I involved my doctors to share my story through my platform. People saw the first story and got mammograms. When I got ovarian cancer, people saw that second story and got genetic testing. Now, with this third story, people in the high-risk community are contacting me. My big thing is vigilance, because if you're at high risk, it can save your life.

It was such a rare thing that I even got this cancer, the current one, after taking every preventive measure available. I needed to get through that psychologically, which was a big deal for me. Since it was so rare, the doctors didn't take this cancer lightly. They were going to be very aggressive with it, and I wanted them to be because I didn't want to take any chances.

I knew chemo was unavoidable. I knew the "red devil" was unavoidable, and Taxol and Cytoxan were unavoidable, as well. I just dove right in and told myself, *This is what I have to do.* Then it was twenty sessions of radiation.

After all of that was through, I was put on Lynparza, a new medication for certain types of cancers [Author's note: primarily BRCA-related cancers]. At the beginning, I was on the fence, so I got three doctor's opinions to make sure they all said the same things. I've been very active in the community, and ten years ago, I sat in on a meeting where Lynparza was being tested. I never imagined that ten years later, I would be taking this drug. As a news reporter who takes pride in knowing things early or even before they happen, actually taking Lynparza was an amazing feeling. I was so grateful to be included in that wave of knowledge, because I know a lot of women aren't. That's why the sisterhood is so important. We do have to really talk about it. I felt blessed to have the Lynparza, and I know everybody can't take it. I got many messages from viewers asking, "Can I take Lynparza?"

when they saw my journey. I don't ever pretend to be a doctor, so I always refer them to their own doctors, but I feel lucky.

It lasts a long time, treatment with Lynparza. I was lucky mine was only a year, but for ovarian cancer, Lynparza is two years. I needed to mark the time, so I would literally stack the pill bottles up each month. I started to build a tower of pill bottles, and it sounds kind of juvenile and hokey, but that was how I did it because I needed to physically see the mountain I was climbing. It helped my brain process the time and it was what I needed to see to feel like I was accomplishing something. Because with Lynparza, you can't measure what you're preventing. You can only see that you're okay and don't have cancer. But there's no sure thing. None of us have the final chapter of our book written.

I keep joking there needs to be a Lynparza support group because even though, yes, I got through it, the whole time I wished that there was more of a network of people who've taken this drug and could answer some questions. When you have your first chemo treatment, you sit in an infusion room, and you have people all around you helping and monitoring everything. The same goes for radiation, they mark you very carefully. When you start your first targeted therapy, in my case, Lynparza, a bag of pills just arrives in the mail. I still have a picture of mine; it had a skull and bones on it for "hazardous drugs." That was so freaking scary. I'm sitting in my house, opening this thing. I mean, am I supposed to wear latex gloves? Do I need a hazmat suit to take this first dose? You're already traumatized enough, having heard your oncologist talk about all the potential side effects of this drug, like a significant blood cell drop, because everything has side effects, and now you're all by yourself. You're not in an infusion room with people all around you checking everything, warning you about everything and saying, "Now, if this happens, here's what we do." There's nothing like that. So, you either have other people you know who've taken it, or you have to figure this out on your own. Honestly, when I saw the hazardous drug warning label I laughed because it was so bad that it was funny, and I thought, *I cannot believe that this is where we're at.*

The Lynparza gave me a horrible taste in my mouth—there's actually a name for it, dysgeusia—that is a bitter taste under your tongue. Not like chemo, it's different. It's very specific, and it sticks with you the whole time. You have a malaise, I called it my "malazy" stage. I felt really lazy and just didn't want to do that much.

The first two weeks represent the adjustment phase for Lynparza. That's when people tend to get nausea. Your body needs a couple of weeks to adjust. If you can just get over the hurdle of the first couple of weeks, your body will say, *Okay, this is what we're dealing with.* That doesn't mean it's great; you'll always have a malaise, but it will get easier, and the nausea will fade. Your immune system is compromised just like with chemo. Basically, Lynparza is chemo light. Regardless, you have to get on with it. For me that meant taking the two pills and going on a run.

When I finished Lynparza, I was concerned about whether I would feel better and how long that would take, which was another big mystery that left me wanting more of a network. I asked a few people in the community, but nobody's just like me. They might be thirty years old and have taken Lynparza and tell me that I should definitely feel better, but that's coming from a thirty-something person, and their case may be totally different.

Now I'm definitely better from the bulk of it. I'm grateful that I was able to keep the high dose and get through the year without much nausea, but I know other people who have struggled with that dose. A bit of advice that was given to me was "just stay on something." Even if you have to reduce the dose, it's better to stay on the medicine than to go off it completely. The last thing I'll say is that when I finished my Lynparza, I experienced mixed feelings about "being done" because finishing treatment can feel like no longer having a security blanket. My advice is to prepare for that.

DR. PORT: Stacey's incredible story, which she has been so generous to share with the world over the many years along the way, exemplifies courage, strength, and incredible optimism. That optimism is part of her fiber, but it's also born of the amazing progress we've made for cases like hers. There have been significant improvements on all fronts of breast cancer treatment, from better technologies for earlier detection to better surgical techniques. But some of the biggest advances have come from the development of newer, better treatments. It's worth reiterating from earlier the number of effective treatments and drugs that have been developed for breast cancer in the last decade equals the number from the prior three decades combined. These therapies, which use a variety of strategies to kill breast cancer cells, have led to the highest overall survival rate ever: 91 percent and climbing. Without going into the details of every single drug, let's review the different classes of drugs and how they work.

IMMUNOTHERAPY

Immunotherapy leverages our bodies' own immune systems to fight cancer. Our bodies are constantly on the lookout for threats: infectious agents, damaged cells that need to be cleaned up, and, yes, cancer cells. Successful cancer cells find a way to evade the body's inherent immune surveillance system effectively enough to grow and spread. In order for immunotherapy to work, the body's own immune system has to be unleashed, and the tumor cells have to generate an immune response. Pembrolizumab (Keytruda) is the main immunotherapy drug for breast cancer treatment.

Excitingly, the triple-negative subtype, which is one of the hardest to treat, seems to generate the most robust immune response. A landmark trial called the KEYNOTE-522 trial showed significantly higher complete response rates (where the tumor melted away completely when given before surgery) with chemotherapy *and* pembrolizumab immunotherapy combined, compared to chemo alone—and translated to significant survival benefit too. When the results from this study were published in 2024, it was a huge development. Now we have something that's most effective in the single subtype of breast cancer where we *used to have* the fewest options. Keytruda is usually administered intravenously, and side effects are mostly related to what happens to the body when the immune system's brakes are removed. The suffix *-itis* means "inflammation," a key component of immune response. So many of the potential side effects of immunotherapy are versions of *-itis*, an overactive immune system affecting one's own organs: thyroiditis and subsequent thyroid dysfunction and hepatitis are two examples. There are more serious side effects that are thankfully rare, like myocarditis and pneumonitis (inflammation of the heart and lungs). Because immunotherapy is often given with chemotherapy, some of the vaguer side effects, like fatigue or nausea, can be hard to attribute solely to this therapy.

PARP INHIBITORS

PARP inhibitors fall into a class of drugs called targeted therapies. Lynparza, the drug that Stacey took, is a type of PARP inhibitor. PARP inhibitors prevent an enzyme called PARP, which repairs DNA damage, from working. The main goal of PARP inhibitors, which target cancer cells, is to prevent them from repairing themselves. As with immunotherapy, PARP inhibitors were a welcome and transformational development. They target cells with BRCA mutations, meaning they specifically benefit another group of patients who are in strong need of improved therapies: BRCA mutation carriers. Unlike immunotherapy, PARP inhibitors, as Stacey described, are taken in pill form. Side effects can include nausea, vomiting, diarrhea, and fatigue, but as with immunotherapy, because PARP inhibitors may be taken along with other medications, we can't always attribute certain side effects specifically and solely to PARP inhibitors. PARP inhibitors can also cause a drop in blood counts, which can then increase the risk of infection (by lowering the white blood cell count), or the risk of bleeding (by lowering platelet count, critical for blood clotting).

OTHER TARGETED THERAPIES

The recent explosion of targeted therapies has advanced our cure rates for other subtypes of breast cancer even further. For hormone-positive, HER2-negative breast cancer, CDK4/6 inhibitors are now the standard of care, in combination with antihormonal therapy. These drugs are called cell cycle inhibitors. While there are a few of them, the ones we use in breast cancer are called ribociclib and abemaciclib. They are primarily recommended by our oncologists in more advanced cases of disease and for patients at a higher risk of recurrence, given factors like their nodal involvement. These medications are also administered in pill form, and side effects include fatigue, nausea, and drop in blood counts, as described for PARP inhibitors above.

BY THE TIME YOU READ THIS BOOK . . .

There will be more breakthrough drugs that improve survival rates for breast cancer. They will continue to advance our quest toward a 100 percent survival rate. That won't mean the same treatment for everyone, but it will mean cures for all.

29

BREAST CANCER FOLLOW-UP AND CAN IT COME BACK?

VIVIANA FIGUEROA, wife, runner, guardian to twins: When you're diagnosed and you're working really closely with your medical team, it's almost like you have a plan, you have a road map, you're given all these appointments and everything's pretty laid out for you. I think the even bigger challenge comes after it is all over. After the appointments start to die down, you're now seeing your oncology team only every three months or every six months. You're seeing your breast surgeon maybe every six months and two years later you go down to only seeing them once a year. Now you're back to counting on your own breast self-exams and things like that.

I think that a lot of people don't talk about the emotional and mental aspect of the challenges you're going to face and the resources that are out there. I know I don't stand alone when I say when it was all over, then I could deal with the mental part. Then I could deal with the emotional part. It's very important to know that it's all normal. It's part of the process. It's okay to be sad or question certain things when treatment is over. You become a little paranoid; when something hurts, you're like, *Oh, it's cancer.* One of my doctors broke it down to me this way, "Once you've gone through cancer, it just means that your medical team is going to be hypervigilant. They're not going to disregard a symptom that you happen to mention, they're not going to disregard something that would appear minor on a scan." That's part of survivorship. It doesn't mean that the cancer is back.

DR. PORT: When so many women get diagnosed with breast cancer, they gear up for it: the fight, the battle, being a warrior. If you spend enough time on any social media site, that's how breast cancer patients rightfully describe themselves. We describe them that way too. More than one of my patients has said to me, "I picked you as my surgeon because you're the one I wanted in the foxhole with me." They knew that I would fight with them and for them with everything I had to offer.

After the initial surgery is over, I see my patients in follow-up for years to come. The first appointment back with me happens three months after their surgery. I go into it with three goals:

1. To see how my patients are feeling, to track how they continue to recover and how the healing process is progressing. Wounds continue to heal, evolve, and mature for months after surgery until the final result at around six months, so the three-month point is a great time to check in.

2. To schedule the next round of imaging. My patients who have a lumpectomy or even a mastectomy on just one side will need future mammograms, sonograms, MRIs, or some combination. At the three-month mark, we schedule that, usually for about six months later.

3. To make sure that my patients have gone on to do any additional treatment that we recommended. By three months, many of them will have started or even completed additional treatments, such as a short course of radiation or chemotherapy.

I use the three-month appointment to make sure no one slips through the cracks. I know it might seem impossible to slip through the cracks with cancer treatment, but you'd be surprised. When I arrange an appointment with an oncologist for the week after our postoperative appointment, the patient almost always follows through. A small percent of patients call to cancel their appointments and either reschedule them or don't. Maybe they don't show up at all for an appointment that we made for them because they had to work, or they couldn't get childcare, or their car broke down, and they couldn't get there. Cancer doesn't come at a convenient time for anyone, and while cancer can get in the way of life, the reverse is also true: life can get in the way of cancer care. When you work in an environment where the patient population is one of the most diverse—not only ethnically and racially but socioeconomically—you see patients for whom managing their life circumstances to include cancer can be overwhelming. Ergo, the three-month visit. Three months, or twelve weeks, is a specific amount of time in cancer care. It is the maximal interval where there is still time to get on board with additional treatment, where we can still deliver treatment safely and effectively, and where we can get a patient who's slipped through the cracks, for whatever reason, back on track. With delays, most of our data shows that the additional treatment ship has sailed. If there is residual cancer, it may have time to regroup and spread. Remember, skipping additional treatment will put some patients at a dramatically higher risk for the cancer to come back. At three months, I want to make 100 percent sure that if a patient decides not to take any additional treatment, she knows what the risks are. I encourage all of my patients, even the reluctant ones, to at least meet with the other specialists and get the information they should have to make their most informed decision.

Then there are patients who don't follow through on the appointments, or who have the appointments but decide against additional treatment. They're scared of the side effects of the treatment that was recommended. They don't want to try a medication after hearing about the side effects their friends with breast cancer experienced. The radiation takes too much time. They would rather take their chances. Can system failure be a reason for patients not getting optimal treatment? Sure, it happens. But when patients don't complete recommended treatment in a timely fashion, it is almost always by choice. For the majority of patients who want to do everything they can to prevent their cancer from coming back, additional treatments are a way of doing something else to fight cancer. They are actively engaged in lowering their own risk while in treatment. While tamoxifen has side effects, many of my patients look at their daily tamoxifen pill as their added insurance policy. Taking that pill is a life-affirming ritual they perform every morning to make sure they are doing everything they can for the best outcome.

As Viviana so beautifully expressed, sometimes, the hardest

part of having breast cancer comes *after* treatment is in the rear-view mirror, after surgery, after chemotherapy, after the dust has settled. My patients go into warrior mode to get through what they think will be the hard part: big operations like bilateral mastectomies or chemotherapy. They put blinders on and put one foot in front of the other, taking it one treatment at a time, plodding through like they would face any other critical life challenge. When that's all done, then what? After treatment, the trips to the hospital are less frequent. So is the communication with doctors. For many patients, the return to a kind of new normal starts setting in. At the three-month visit and the visits to follow, some patients take on a "deer-in-the-headlights" look that's totally different from the one they might have had at our first meeting, when they'd just been diagnosed. That's because the transition from actively fighting cancer to moving past treatment can be fraught with its own complicated feelings. In this post-treatment phase, the *What do I do now?* conversation often comes up. In active treatment, there is a *feeling* of doing something. When treatment ends, so does the feeling. That's when the worry about cancer coming back often begins. Many patients voice the feeling that after treatment is over, they feel like they're just sitting around, waiting for the cancer to come back. For patients, this is an important conversation to have. For me, it's the final purpose of my three-month follow-up visit, although it can come later for patients on a longer course of chemotherapy.

Can cancer come back? Yes, it can. And here's where if you are newly diagnosed, or going through treatment, you may want to stop reading this chapter and skip to the next. For women who have a lumpectomy, the cancer can come back in the same breast, but the risk is quite small: less than a few percent. A new breast cancer can also develop in the other breast for those who did not remove it. That, too, has an extremely small risk, unless the patient is BRCA positive. But most ominously, breast cancer can come back in the body, as metastatic disease: stage IV.

Given all the treatments and new developments we have, the chances of this occurring are extremely low for most patients. For many women with early-stage disease, survival rates are above 91 percent. But this percentage can go down—and the risk for recurrence can go up—for people who choose to opt out of certain types of therapy.

Cancer brings up control issues. For many people, a cancer diagnosis is a major reckoning with the fact that there are serious things in life we can't control. That's why so many of my patients come in after treatment saying, "What more can I *do* to make sure my cancer doesn't come back?" I tell them that while we don't have control over every aspect of their outcome, we *do* have control over many parts of it. This is the point where I can discuss lifestyle changes with my patient, if they're relevant and the person is ready.

For most of my patients who have completed treatment, I say, "You have done everything you were supposed to do!" I don't have a crystal ball. I can't see the future for myself, much less my patients. What I can do is tell my patients a story about outcomes. I tell them that I have the most incredible husband: a smart, funny, good-looking, brilliant surgeon. He has one major flaw: his driving. He drives too aggressively in my opinion. Pretty much the only way I can get him to be a bit more careful behind the wheel is by reminding him that our two beloved dogs, one of whom is usually riding shotgun on my lap, would not survive the impact if he crashes, since airbags are designed to protect people and not twenty-five-pound animals. When I get in the car and go anywhere, with or without him, how do I know I'll make it from point A to point B? The truth is that the possibility of getting into a life-threatening or, God forbid, life-ending car crash is extremely low. But it's never zero. The same goes for crossing the street. Nothing we do is completely risk free. So what do I do when I drive? I wear my seat belt. I don't text and drive, and I certainly don't drink and drive. When he is driving, I frequently tell him to take a chill pill. These are the things my husband and I can actively do to prevent ourselves from having a bad outcome while driving. No matter how safely we drive, we can't control if some other idiot is texting his girlfriend going eighty miles per hour and hits us. We can never de-risk a situation down to zero, and the tiny risk that remains is out of our hands. Plus, if we stopped and thought about the risks of every single thing we do every day, we would be paralyzed! We would never leave our homes or have lives worth keeping ourselves safe to live.

Cancer recurrence risk is just like driving. You wear your chemo seat belt, if you are told to. You have the right operation for your case. You take antihormonal medication, if it's indicated to do so. You have radiation, if it's recommended. Then, you have to live your life knowing that you did all the right things and that you have the highest chance of survival you can possibly achieve. My job is to help my patients get to the sense of freedom and peace that comes from knowing you have done everything within your power to keep yourself healthy—and let them move forward with totally appropriate and well-informed optimism.

Many of my patients' worlds are rocked when they hear about a bad outcome from breast cancer, whether it's someone they know personally, a friend of a friend, or even a celebrity. Given how common breast cancer is, of course we hear about bad outcomes. A story in the news about a famous person dying of breast cancer can spur dozens of calls from patients worried about themselves. They ask, "How did this happen?" and, more ominously, "Could it happen to me?" "Of course," I tell them. "Anything can happen to anybody. But," I continue, "what if instead of asking 'Could this happen to me?' we shifted to asking, 'Did that person do everything they were recommended to do?'" I have patients who choose not to take recommended chemotherapy or don't go on the recommended antihormonal therapy. Or they try it and stop. These are all personal choices, but when these stories about cancer coming back circulate, I have never once heard the person admit that they didn't follow the recommendations for treatment—and that maybe if they had, they would not have had a recurrence. There are 40,000 estimated deaths related to breast cancer each year. That's far too many. We know that 30 to 50 percent of those bad outcomes are related to forgoing certain aspects of treatment. In other words, many of those deaths are probably preventable. Can breast cancer come back even when you do all the right things? Of course, just like a person can die in a car accident driving safely, wearing a seat belt, and paying attention. The risks are low for both of these outcomes. As I've said earlier, I tell my patients that I'm a cancer doctor, not the cancer police. No one comes to your house to drag you to chemotherapy or makes sure that you take a pill every day. People have to make the choices to do those things as individuals. That's where someone else's experience, outcome, or even advice should end before it applies to you. As I also tell my patients, their scary story is not your story.

THE BREAST TREATMENT ADVICE

TREATMENT ADVICE FROM MY PATIENTS

GINA S., medical professional: People always ask me about working during chemo. I'm someone who loves structure in my day, and I knew if I wasn't going to be able to work, that would not be good for me. I did have a conversation with my team about how I might not be at 100 percent with my work. They were so understanding about it. They said, "Whatever you need. If you need to take a nap during the day, just let us know. Ping us, 'I need to step away for an hour.'"

In the end, I was able to work three days a week for the twelve weeks I had chemo. I decided to take three days off to recover. I scheduled my chemo on Thursdays. I felt okay on Fridays, but I told myself, *I'm not using this day to go to work. I'm using this day to get outside and go for a walk or go get a cup of coffee somewhere before I feel like crap for the rest of the weekend.* By Monday or Tuesday, I was okay again and ready to go back to work. I sacrificed my weekends to the chemo gods so I could be mostly functional during the beginning of the workweek. I know that's not the answer for everybody, but it worked for me and kept some of the normal structure in my life that I really needed.

KATHERINE ESKOVITZ, lawyer, author, and mom of three: When I was first diagnosed, everyone seemed to know someone who'd had breast cancer, and everyone had a list of "must-see" doctors. I was still absorbing the news when my well-meaning mother-in-law, a survivor of a grueling year of chemotherapy, texted: "By the way, I had the same kind of cancer as you. If you need chemo, get the ice treatment for your hair." That sent me into a downward spiral. I wasn't ready to lose my hair. I wasn't ready to consider chemo. Another friend texted me a vitamin and supplement regimen, fasting retreats, and a link to leaders of alternative cancer treatments in Mexico. I was not prepared for the onslaught of information, but I appreciated it. I filed it. I considered all of it and made notes to ask my doctor. Breast cancer had so many different shades and colors: a lumpectomy followed by five days of radiation; six surgeries and a double mastectomy; chemo, radiation, and five years of medication. I didn't yet understand that every diagnosis is different and that everyone has an opinion about it. As the days passed, I started to adjust. In the many, many texts and conversations I had with friends and family, I absorbed some very wise advice: take it one step at a time, ask for what you need, and don't worry about others when you need to take care of yourself. I forced myself to get past the terror of the word *cancer*. I filtered information, set boundaries around unhelpful calls and texts, and—most importantly—chose a medical team I trusted and listened to them.

AMANDA KATZ, certified clinical medical assistant: Honestly, sometimes I feel like an impostor because I don't have such a traumatic story of cancer. I was able to get over it a little faster and separate it from the rest of my life and it hasn't defined me. It's this weird psychological battle almost like I have survivor's guilt. I had actual cancer and yes, I had to do something extreme. I did it because I was operating from the mindset of *I don't want this back in my life*. I also didn't want to do one side and be doing mammograms and worrying about the other side for the rest of my life. But there still is a lot of impostor syndrome because it feels like I didn't suffer enough. I didn't go through the yearlong process of making implants look like my natural breasts. It's an *I don't belong in this club* kind of thought.

SARRAH STRIMEL BENTLEY, advocate, motivational speaker, former Broadway actress, and mama of Chance: I saw something online which really bothered me recently. It was a doctor talking about someone who didn't do traditional treatment for breast cancer like chemotherapy. She did alternative healing and her tumor was gone. Now, I'm a yoga teacher, but it's just so dangerous to say these things. Because if you want to bring in other therapies and you want to find an integrative oncologist, amazing, but there is a reason why we have the course of treatments that we do. It's because they work. I wanted a doctor who leads with science and then I could do the woo-woo stuff on the side. And that's also true if you don't have cancer, even for general health.

EMMA B., insurance agent: I chose to have my treatment near my apartment so I could walk back and forth, which I did often. I loved window-shopping as I walked the few blocks home. Even if I was tired, it felt good to get some fresh air and clear my head after sitting in the chemo chair for four hours. I would wander into the stores every now and then, and definitely engaged in retail therapy. I bought a pair of new sandals, since my feet seemed like they were starting to swell from the chemo and the hot weather. I bought a blouse that I knew might not fit after all of my treatment but was so beautiful that I wanted to wear it now. I would find myself buying the oddest things. I bought a pair of candlesticks at this home store. I had no use for new candlesticks. I bought these crazy expensive designer sunglasses. I never wore sunglasses like that, but that summer, I thought they made me look glamorous, even with the wig and the chemo weight gain.

When I got through chemotherapy, my husband bought me this necklace with a tiny pink sapphire in the middle and a small diamond on both sides, one for each of my kids. I never would have bought it for myself, but I love it, and I wear it proudly. I know it's just a *thing*, but it's a thing that means the world to me. And while it might be weird to get a reward or a "trophy" for finishing chemo, I think of it as the chemo version of a "push present." And it's a reminder that I survived thanks to the support of my doctors, my nurses, and everyone I love. That's worth more than all of the things combined.

MY ADVICE

When it comes to treatment for breast cancer there is truly no one-size-fits-all, since the options for treatment have gotten better, more varied, and more refined over the years. Seek out a team with specialized expertise in breast cancer treatments, so you know what is recommended for you is tailored and appropriate for your specific case. Follow the recommendations; don't let fear lead to avoidance of appointments or treatments. Expect to need rest and listen to your body. Always seek support when you need it. And remember, as you make your way, whether it's chemo or no chemo, antihormonal treatment, or radiation, you will make it through.

SPECIAL SITUATIONS

30

RARE TYPES OF BREAST CANCER

ELIZABETH A., driving instructor: I was fifty-two years old when I noticed a rash on my breast, and it seemed kind of swollen and heavy. At first, my doctors thought I had mastitis—an infection—but I was a little old for that, plus I wasn't breastfeeding, which is how mastitis usually develops. Finally, after a month of it not going away while on antibiotics for the "infection," I went back to my doctor, and more tests were done. I had a mammogram, and a sonogram, and then a biopsy and it turns out I had a rare form of breast cancer called inflammatory breast cancer. It was extremely scary, because everything I read about it was not good: it's stage III, it's harder to cure, and all that. Plus, the treatment plan for inflammatory breast cancer is really different from the plans that some of my other friends had, much more intense and extreme. Some of my friends got radiation, but not chemo. Some had surgery but no radiation. I got it all. That was three years ago, and while I have been through a lot, I'm still here and I plan to stay.

DR. PORT: Let's go over some of the rarer types of breast cancer, the types that make up less than 5 percent of all new cases. Some of these are first detected based on physical exam findings, which should never be dismissed. So, while screening is always important, physical exam changes are also critical to be aware of even when recent imaging test results may have been normal. Again, I want to emphasize the importance of knowing what is normal for you as it relates to your own breast exam, so subtle changes might be more easily identified.

INFLAMMATORY BREAST CANCER

Approximately 1 percent of all breast cancer is this type. While inflammatory breast cancer can be associated with a lump or enlarged lymph nodes, the specific findings that distinguish inflammatory breast cancer are the characteristic skin changes. The skin can be thickened and reddened, appearing like an infection, and often misinterpreted as one. The reason for the skin changes is cancer involvement of the skin lymphatic channels, blocking their drainage, causing swelling and skin pitting. This sign is called *peau d'orange*, French for "orange peel," because that's the appearance that the breast skin takes on. To make the diagnosis, one needs only to see these skin changes and have a biopsy of any part of the breast that shows breast cancer. We often perform small biopsies of the breast skin to definitively determine the diagnosis, but these biopsies are only confirmatory, and the diagnosis of inflammatory breast cancer is based on what we see and feel on examination. Inflammatory breast cancer is more aggressive and thus requires more aggressive and extensive treatment. It is staged as stage IIIB, so quite advanced.

The treatment options and cure rates for inflammatory breast cancer, just like all breast cancer, have come a long way. The treatment pathway for inflammatory breast cancer always involves chemotherapy first because with the extensive skin involvement, there is almost no way to remove all the disease with up-front surgery. Chemotherapy is often successful in shrinking down disease, including in the skin, allowing for successful surgery to follow. During chemotherapy, I always have my patients come back a few times during their course of treatment, and we can often see the skin changes and underlying swelling and masses resolve along the way. At the conclusion of treatment, surgery can be performed, and it is often a modified radical mastectomy, given the extent of original disease and the fact that lymph nodes are frequently involved. Surgery is then followed by radiation. The chemotherapy, followed by surgery, followed by radiation approach is extensive and may seem possibly excessive in this age of de-escalation. But the aggressiveness of this cancer in particular must be matched with an aggressiveness in treatment to potentially achieve cure.

Having reconstruction with inflammatory breast cancer can be controversial. On the one hand, we need to preserve as much skin as possible to allow for reconstruction. On the other hand, given the extensive involvement of the actual breast skin, we try to remove as much of it as possible. These opposed goals can make decision-making for reconstruction complicated, and most commonly, it is recommended to delay reconstruction until after the entire course of treatment—chemo, surgery, and radiation—is all completed. Because of the radiation to the skin and chest wall, implant-based reconstruction to follow is rarely achievable, so most women ultimately have tissue-based reconstruction.

PAGET'S DISEASE

Paget's disease is another rare form of breast cancer, comprising approximately 1 percent of all new cases. It is also most frequently identified by physical exam findings, and often by a patient. Paget's disease starts in the nipple and as such, a woman might complain of a "rash" or scabbing on her nipple. Some women say they are having nipple discharge, and find staining on their bras or pajamas, but it's usually not true discharge from the nipple, rather it's oozing from the nipple surface, which can be raw or excoriated. Often attributing this new finding to irritation from something they wore or to eczema or some other skin condition, patients will often see their primary care doctors or even a dermatologist at first, who might prescribe topical medications, like steroid creams or antibiotics, to be applied to the area. With cancer, these topical treatments will not help, and it won't go away. So, while it's reasonable to try these remedies for a week or two, if there is no resolution, seeing a breast surgeon is the best next step. In some cases, breast imaging studies, a new mammogram or sonogram, might find concomitant concerns within the breast, like a mass behind the nipple or calcifications. Normal imaging does not definitively eliminate the possibility of Paget's disease, and many with this disease do have otherwise normal imaging findings. A small biopsy of the nipple is necessary to diagnose Paget's, and then the treatment and cure pathways are similar to that of standard breast cancer. Because the nipple is involved, it has to be removed as part of the surgery. This can mean a mastectomy for some women, or a specific type of lumpectomy removing the central portion of the breast including the overlying nipple, which can also be done with limited disease within the rest of the breast. As discussed previously, this type of lumpectomy is called a central quadrantectomy.

OCCULT BREAST CANCER

Less than 1 percent of breast cancers are occult, which literally means "hidden." With occult breast cancer, the first thing that is discovered is not a lump in the breast or any change on imaging. It starts with finding an enlarged lymph node under the arm, usually by a physical exam. A woman might be washing herself in the shower and notice a lump in her armpit. Or a doctor may be performing a standard physical exam and find it that way. With a suspicious lymph node, we usually use ultrasound to zero in on that node and then perform a needle biopsy to make the diagnosis. While a needle biopsy of a concerning lymph node can be normal or even show another type of cancer, such as lymphoma, it can also show breast cancer. When the biopsy shows breast cancer, it means that there *is* a cancer somewhere in that breast, but it's either so small or so hidden that we can't find it by exam, mammogram, or ultrasound, and the first sign of it was this spread to lymph nodes under the arm. Importantly, occult cancer can be an invasive ductal cancer or an invasive lobular cancer, or any breast cancer cell type really. *Occult* simply refers to the fact that the primary site in the breast is hidden, and refers to the way that this rare cancer presents. With occult breast cancer, we always perform an MRI, and approximately 60 percent of the time, we will find the source cancer in the breast. Then, treatment can proceed as usual. The tricky part is what to do when, even with an MRI, a cancer in the actual breast, what we call the primary, can't be found. When this happens, our options are either to perform a mastectomy, assuming there is a cancer in there *somewhere*, or to treat the breast with radiation. During surgery for occult cancer, we also usually perform an axillary dissection, given that nodes are usually extensively involved since that is how the cancer was found in the first place.

MUCINOUS, TUBULAR, AND MEDULLARY

All three of these are rare subtypes of breast cancer that usually are slow to spread and are often associated with better prognosis than ductal and lobular cancers. They are essentially treated the same as the more common subtypes, but given their low likelihood of spread or recurrence, checking lymph nodes with smaller lesions can often be avoided as can more aggressive medical treatments. Medullary carcinomas can be triple negative but are often associated with a better outcome than standard triple-negative breast cancers.

31

YOUNG WOMEN WITH BREAST CANCER

HOW IS IT DIFFERENT?

GINA S., media professional: I was thirty-two when I got diagnosed. I had what I think is a very healthy lifestyle: I wasn't a heavy drinker, I didn't party much, I was a marathon runner.

Breast cancer was the last thing on my mind. I was going through a really tough time emotionally because of a particularly bad breakup. I'm someone who keeps up-to-date with my medical appointments. Every January I have my annual ob-gyn appointment.

This particular January, I had to push it because of the breakup. I was in between apartments in New York and had moved back in with my parents in Massachusetts. I'm so thankful I pushed it because I ended up going when I moved back to New York in February. When my doctor did the breast exam, she found the lump, and said, "We need to get this checked out."

My mother had DCIS and a lumpectomy when she was fifty. That was over twenty years ago with no recurrence. My grandmother also had ovarian cancer and unfortunately passed away with that at seventy-seven years old.

Considering my family history and the fact that I have dense breasts, my ob-gyn said I needed a mammogram and an ultrasound. When I was getting the ultrasound, the tech was taking a long time and I was like, "Oh, what's going on?"

I kept my peace, but when the radiologist came in and said there were some spots that needed biopsies, I completely froze. That was the last thing I was expecting.

I did start to cry. It was a Friday at the end of the day, and I knew I wasn't going to be getting in until the following week to get these biopsies. I'm very much an anxious person, and with everything I went through with the breakup just a few months before that, it was a perfect storm. I called my mom crying and she reassured me, "They're just doing their job. If you need me to come down from Boston, I'll be with you." I told her that I could do it; I would treat it like just another appointment. I had to get three biopsies. You hear the word *biopsy* and you think it's going to be the scariest thing in the world. It was not. It's nothing that someone wants to go through, but the level of care I received, and the compassion, made it okay.

Then came the worst part: the waiting game. This is when I had no resources and I went to, as I'm sure a lot of patients do, Reddit and Google, and I doomscrolled.

It was a numbers game in my head: 50 percent chance

of coming back as nothing. *You're probably gonna be fine.* But since I'm very anxious and I had three biopsies, my mind went to the dark part and I thought of course, *One of them is gonna come back bad.* So, it was a lot of doomscrolling, a lot of, *Oh my God, I'm gonna need chemo. I'm gonna lose all my hair. I'm going to be single forever. Holy crap, what is a chemo port?* I was spiraling.

When I got the call that I did have cancer, it was an out-of-body experience. I felt like I was going to throw up. I cried hysterically. I remember texting my mom just, *I have cancer.* I didn't know what else to say. Texting friends, *I have cancer.*

It's a weird feeling as a thirty-two-year-old who felt like her life was about to take off again after moving back to the city after spending some time with my parents in Massachusetts. Where was my single-girl-in-New-York era?

My mom came down to come to the appointment with me. As soon as we had a very long and extensive conversation with Dr. Port about what the early stages would look like, I felt at peace that there was a plan and glad to be out of the in-between stage of doomscrolling. Dr. Port reassured me that this was curable. Obviously, I think I can speak for many women when I say that's the number one concern.

MELISSA D., graphic artist: Breast cancer never comes at a convenient time for anyone. I am sure of that. "This really is happening at a point for me when there's nothing going on in my life," said no one ever. People have plans. They have lives. I had a dress. A wedding dress to be specific. I went in for my last fitting for my amazing wedding gown. It's strapless, with a sweetheart neckline. As I was "pouring" myself into it and holding it up in front while getting zipped up in back, my hand brushed against my left breast, and I felt a lump. I moved my hand, pretending to adjust my dress, and, yup, there was definitely a lump there. A lump I did not feel the day before. I said nothing, not wanting to alarm my overjoyed mother waiting outside to see the dress in its final state. I put on my happy face, stepped outside the dressing room, and watched everyone beam. Inside, I was frantic. I kept trying to calm myself down, saying maybe it was nothing. But it wasn't nothing. I saw my doctor two days later and had a mammogram and a sonogram, then a biopsy. I was thirty years old. It was four weeks before my wedding, and I was diagnosed with breast cancer.

Since being diagnosed, I have met so many young women with breast cancer in my age group in chat rooms, in person, and in support groups. Many were diagnosed after waiting months to see if the lump went away, or worse: they sought medical attention and were told not to worry and were sent away with no further testing—or were actually told they were too young to get breast cancer! Even though the timing of my diagnosis was the worst possible, I didn't blow off my concern or wait until after my wedding. My advice is to make sure everyone knows three things. You are *never* too young to get breast cancer. Don't let anyone tell you otherwise. Go to a center of excellence for breast cancer—they exist all over the country—where the team has taken care of enough women in our age group with breast cancer to know that we have specific concerns that need to be valued and addressed, like future childbearing, breastfeeding, and genetic factors. Lastly, we still have the potential to live long healthy lives. While we should never be going through this at our age, there is so much room for optimism.

DR. PORT: In recent years, the incidence of breast cancer in young women has increased. A recent study showed an approximate 20 percent rise in breast cancer rates among women younger than forty, and the increased risk was more pronounced among Black women. There is no clear understanding or reason as to why this is the case, but there are a number of theories. First of all, breast cancer in younger women is more likely to be related to a family history or genetic predisposition compared to breast cancer in older women. Among *all* women with breast cancer, only approximately 5 to 6 percent carry a BRCA mutation. Among women under forty-five with breast cancer, that rate is double: approximately 12 percent. That still leaves a large proportion of younger women who get breast cancer without a family history or genetic predisposition.

What are the other potential causes or risk factors? Almost all of the factors that we can identify that might be at play relate, directly or indirectly, to hormones. The first factor is increasing obesity levels in all young people. The obesity rate is currently approximately 40 percent in American adults. Just as important, the obesity rate has climbed in younger populations, leading to our young people's bodies being defined as obese at younger ages. In adolescents, twelve to nineteen years old, the obesity rate is currently above 20 percent. For children as young as six to eleven, obesity rates are nearly as high. How does obesity impact the development of breast cancer? The short answer is the connection between fat stores and hormones. In young women, the ovaries are the main source of hormone production, but fat stores also can produce hormones, adding to the body's supply and circulating levels. Most breast cancers are hormonally driven and fed, so it stands to reason that increased hormone levels that start earlier and last longer could increase the risk of breast cancer. When that increased level starts at such a young age, sometimes even before puberty, the overall exposure to higher estrogen levels over years of childhood, adolescence, and young adulthood adds up, and could potentially, we think, contribute to an increased risk.

Independent of the higher levels of circulating estrogen from fat stores, there may be another obesity and hormone connection at play. Remember the chapter on hormones? Young women and girls are menstruating earlier and earlier. Between 1950 and 1969 the average age of a girl's first period was 12.5. Thirty years later, between 2000 and 2005, it decreased to 11.9 years old. While this might not seem like a huge change, the number of young girls experiencing extremely early onset of menstruation, meaning before the age of nine, more than doubled from 0.6 percent to 1.5 percent. Earlier onset of menstrual cycling is also a risk factor for breast cancer, and it may be that larger amounts of circulating estrogen generated by fat stores may indirectly increase risk for breast cancer by causing cycling to kick in at younger and younger ages.

Another hormone-related factor that increases risk for breast cancer is a lack of childbearing or delaying childbearing, which is increasingly what women are choosing to do. A recent census study showed a decrease in childbearing among women in their twenties and an increase among women in their late thirties and forties. Childbearing is theorized to decrease breast cancer risk by interrupting menstrual cycles, even though hormone levels are consistently high during pregnancy. Finally, long-term use of birth control pills can marginally increase the risk of breast cancer, and the CDC estimates that approximately 14 percent of women between the ages of fifteen and forty-nine are on the pill. Birth control pills can have protective effects against other far more lethal cancers, such as ovarian and colon cancer, so increased risk of breast cancer may be canceled out by those other benefits, depending on the individual and her particular underlying risks or concerns.

The role that our environment plays in the increase of breast cancer among young women remains unclear. Do the hormones cows are pumped with end up increasing hormones our children are exposed to through the milk they drink? What about endocrine disruptors? They're a class of chemicals that can mimic hormones or block them and are found in everything from cosmetic products to pesticides to nonstick pans to Tupperware. They've been implicated in earlier onset of breast cancer. With environmental factors such as these, causality is difficult to prove, so there is no evidence that any of these individual factors are at play. Furthermore, eliminating exposure to all of these potential culprits would be almost impossible.

Although we know that breast cancer in young women has increased, the putative causes include many factors that are theoretical at best. Hopefully, ongoing and future research will show us more definitively how factors that we can control—and even those we can't—influence the risk of developing breast cancer at a young age. More importantly, it can hopefully offer clear-cut pathways to stave off the trend.

Breast cancer in young women, no matter how you define young, almost always comes as a surprise and is always heartbreaking. Thankfully, with specialized, personalized care and a healthy dose of optimism, we get our patients through it and give them the best possible chance for a cure, so they can live out the rest of their lives and fulfill their dreams.

32

FERTILITY PRESERVATION

KEEPING ALL THE OPTIONS OPEN

SARRAH STRIMEL BENTLEY, advocate, motivational speaker, former Broadway actress, and mama of Chance: Nine months before I was diagnosed with breast cancer, I met the love of my life in New York City. We would talk about the kids we would have. James, my now husband, would say, "Our daughter's going to have crazy curly hair like you, and our son is going to be quiet and reserved like me." Then I got diagnosed with breast cancer. I realized I may have to lose my breasts. I may have to lose my hair. I may have to go through chemotherapy. I may have to be in pain. I had no idea that certain treatments, not all, but certain treatments for breast cancer could affect a woman's fertility.

That very much changed how I viewed my breast cancer diagnosis because due to its specific type (ER-positive, PR-positive, HER2-negative early breast cancer), my treatment was definitely going to impact my fertility. That's when my doctor gave me the opportunity to do two rounds of IVF before I started chemotherapy. She saw how in love we were and knew how much we wanted a family. My medical team gave me the time and opportunity to create what we'll call an insurance policy because nothing is certain. We had to decide if we wanted to freeze just my eggs or our embryos together. Embryos do have a higher chance of viability. We weren't married, but we were in love, and this was our only shot. We decided to freeze embryos, and we did this before treatment, to keep my options open.

After my chemotherapy and radiation treatment, I was put on tamoxifen, which would be for five years, maybe longer. My oncologist also said, "Considering your age, your cancer subtype, and the medications we have you on, I don't feel comfortable giving you a break from the medication to carry this one embryo. You can't become pregnant yourself on tamoxifen, because it causes birth defects. So, I think you need a surrogate." That was another huge ninety-degree switch in my journey.

It was nothing I ever saw coming from a family-planning standpoint. However, it was an experience that I was able to find immense joy and immense gratification in. I now have my miracle baby. We even got really close with our surrogate and her family.

There is a life after breast cancer, and it's an awesome life. Do I forget that I went through it? No. I honor it. While it remains a part of me, there are some days I don't even think about it. You have to get through it to get to it, but there is another side.

DR. PORT: As Melissa mentioned in the last chapter and Sarrah pointed out here, young women with breast cancer have unique considerations that older women don't. Most of them are related to potential future childbearing. Here are the main factors to consider:

1. Young women with breast cancer more frequently need chemotherapy as part of their treatment regimen. Chemotherapy can render a woman infertile by damaging the ovaries and the eggs that live inside them, thereby preventing her from becoming pregnant in the future. As a result, we often encourage women to go through a procedure to harvest and freeze their eggs, prior to initiation of chemotherapy, to preserve the possibility of childbearing, if they might want that in the future. A woman can potentially undergo in vitro fertilization (IVF), if it is safe after breast cancer treatment, and bear children herself. Before the 2020s, there was very little insurance coverage offered for egg preservation and IVF in the US. Each cycle could cost between $20,000 and $30,000, and it often takes more than one cycle to become pregnant. Without insurance coverage, the costs are prohibitive, and as a result, the process was only accessible to a small segment of the population who could afford it. Now, eleven states have laws requiring insurance coverage for both egg harvesting and IVF. Other states have laws about covering one or the other. Obviously, you need to have both to have a baby. Still, there's progress being made in making the process for future childbearing significantly more accessible to a larger segment of the population.

2. The second factor that can affect childbearing is the use of antihormonal treatments. Tamoxifen is often recommended to young women as a hormone blocker to prevent breast cancer recurrence. Patients take it for at least a five-year

course, sometimes ten. While women *can* get pregnant on tamoxifen, they absolutely should not, since the medicine causes birth defects. Here's the dilemma: you need to take tamoxifen for five years after surgery but can't get pregnant on it. Often for young women, that five-year period represents the time they planned to get pregnant. However, by *not* taking tamoxifen, a woman potentially increases her risk of recurrence and compromises her overall well-being. Those are two terrible options for a woman who wants to get pregnant. The good news is that we have some relatively new data to show that it may be safe to interrupt the five years of tamoxifen to make time for a pregnancy in some women. In those cases, a woman might take tamoxifen for two years, stop the medicine and let it wash out of her system for a month or two, get pregnant, and then resume tamoxifen for the final three years after giving birth. Of course, getting pregnant does not always happen immediately, and there can be other challenges. Another option that does not involve any disruption to treatment is surrogacy. Using a surrogate is not legal in all states and can be expensive, but I have had many patients have their own biological children this way. The point is that where there is a will, there is often a way. For a woman committed to having a family, we, as doctors, must be willing to work to help her achieve that goal. We get proactive about sending our patients for fertility preservation consultations prior to initiating therapy, so no potential baby bridges get burned.

3. Breastfeeding after breast cancer can be challenging, although it's not impossible in all cases. Having a bilateral mastectomy precludes breastfeeding. Removing all of the breast tissue eliminates milk production and delivery. A single

mastectomy preserves the ability to breastfeed in the other breast. After a lumpectomy, women are often unable to breastfeed from the cancer side, even though they still have that breast. Radiation, often given after a lumpectomy, can affect the ability to breastfeed. Similarly, breastfeeding from the unaffected breast should not be a problem.

If you are of childbearing age and diagnosed with breast cancer, make sure that you know all of your options. Talk to your doctors to find out if pregnancy after breast cancer is safe and achievable for you and if you are capable of delaying treatment to optimize fertility preservation. It's not wise to delay treatment significantly, and please understand that pregnancy after breast cancer is not always possible for everyone in every way. If you are forty years old, it's hard to get pregnant and carry a healthy pregnancy even without considering a breast cancer diagnosis. Taking a beat to explore all of these options *up front* before any treatment gets delivered is critical, so you don't cut off any realistic options. This process involves coordination between your breast surgeon, your oncologist, and a specialist in fertility preservation. Most top-tier institutions have all three and will coordinate for you. If you start somewhere that doesn't, my advice is to get to a place that does and will support you in the way that you need to be supported.

DISPARITIES IN BREAST CANCER CARE (AND HEALTH CARE IN GENERAL)

CHIQUITA CARTER, retired field examiner: I joined the military at sixteen with delayed entry while I was in high school because, even though I had a college scholarship, I needed the discipline. Right out of high school, I went to basic and AIT (advanced individual training). Then, I went straight to war. I went to Desert Storm and Saudi Arabia, where I was exposed to a lot of stuff, like chemicals and burn pits. I was in several SCUD attacks and did get exposed to biological weapons. We had to take nerve agent pills. They made me break out in these rashes on my breast. That's when the whole thing started with my breasts. After that, things just kept popping up. I was always wondering, *What's going on, and how is this going to manifest later in life?* I was really young, but also I had to consider that my maternal grandmother was diagnosed with breast cancer and passed away from it in the '60s, when she was thirty-three years old. I always feared that I would get it too.

I had a scare when I was thirty-eight, and they said, "Because you're high risk, you need to start getting tested." Fast-forward to age forty-six. My mom had just passed away from lung cancer when I got diagnosed with breast cancer. It was really scary. I felt a lump in the upper part of my left breast. Since it started out with just pain and I was in my midforties, I thought it could be menopause, so I left it for months until I actually felt a lump.

Whether you have certain risk factors or not, no matter what your race is, you should get screened. I didn't know that Black women are more susceptible to triple-negative breast cancer. I try to educate people on getting mammograms and doing breast exams, especially now that many people can get mammograms almost for free.

My then-fiancé, now husband, told me to get the lump checked out. I felt lucky to have insurance because I was a veteran. My primary care provider felt the lump and referred me to the place that diagnosed me. I was devastated. I felt my life was over. When they first caught it, we thought it was stage I, but they found out it was actually stage II. Everything started moving really fast after that. I got a lumpectomy, and they removed four lymph nodes. I ended up getting radiation and then AC chemo with the red devil. I couldn't complete it because I got really sick, and my neuropathy was really bad. I was in the hospital for a month. At first, they thought I had sepsis or pneumonia. I suffered until they finally called a pulmonologist who diagnosed me with radiation pneumonitis. They gave me steroids, and I started feeling better, but they made me get really fat for the eleven months I was on them.

Since I couldn't complete the full treatment, I blamed

myself when the second round of cancer came. The first doctor kept saying, "You have to complete it, or we won't get it all." I thought, *Maybe I should have just gotten a double mastectomy.* I was getting yearly mammograms after my first cancer, and I was a little late to getting my mammogram in 2023. That's when I got a new cancer on the other side. I wanted to get a double mastectomy this time to get it over with, but my doctor at the time only wanted to do the right breast. I told her, "I had it in the left breast before, I want to do both." I saw Dr. Port, and while she confirmed that even with a double mastectomy, the cancer could still come back, we did both.

Even though I was devastated and really thought I was going to die the second time, I was blessed both times with the outcome. I know people who have died from it. Both cancer diagnoses were very traumatic. It was so hard to get through chemo the second time. I had to keep getting blood transfusions. I could barely do anything. My neuropathy got worse than before, and the lymphedema was worse than before. You feel like, *Okay, I'm already not what I used to be, then I get cancer again and my appearance changes.* The weight gain, your skin and nails look dirty, you lose all your hair. All of that all over again.

A lot of us Black women have always felt like we didn't get equal treatment, when it came to screening and medical care. A lot of us are afraid to go to the hospital for routine checkups because a lot of us don't have medical insurance. Even if we do have Medicaid, not a lot of care really gets covered. I've talked to other women who have that experience. We feel that we aren't taken seriously, that doctors think we are exaggerating. "We're strong Black women, we'll be fine." We hear that a lot. Many of us die from things because we're not taken seriously. At the same time, the fear is, *What if they do find something? Am I really gonna be able to fight it? Are they really gonna have my back?* It's not that we don't care about getting screened; it's that we don't get treated fairly when we do.

When it comes to research, a lot of us Black women feel like we're guinea pigs because they're just now coming up with more research on how breast cancer affects us. I've been in groups with women from other races, and it seems like they often have an easier time. To me it seems like radiation skin issues might be more common for Black women. When I had radiation, I got really burned. My sense is that treatment might affect us worse. People will say it's statistics. Only 30 percent will get neuropathy. Only 30 percent will get lymphedema. But it seems like the 30 percent is always Black women. The treatments just aren't tailored to us.

In the army, they told us to suck it up and drive on. That was my motto with breast cancer. In the army, there's discipline, hard work, drill sergeants, and drop privates. You're in a situation where you don't know if you're going to live to see the next day. You're getting attacked. You see people with gunshot wounds. It was tough, but it made me want to live. You just have to keep pushing, and I'm glad I did. When I went into the military, I didn't really have much hope. I didn't see myself being where I am now. During treatment, I kept thinking that there's got to be some light at the end of the tunnel.

Honestly, I still don't totally trust the medical community because I've had my share of bad experiences, but I feel like there's somebody who's got it worse than me. Some people don't survive it, but I am a survivor, so I'm blessed.

DR. PORT: There are significant disparities in breast cancer outcomes across women of different ethnicities, races, and sociodemographics. Those disparities are, in part, related to access to care and health care system factors that we see more broadly across all health care, not just breast cancer. The pandemic and its fallout, for example, taught us a lot about health care disparities and put the system's glaring inequities and discrepancies front and center. There were heroic efforts from health care workers and incredible technology that saved thousands of lives, but hospitalization and death rates were not the same across all demographic groups, as I witnessed myself when elective breast cancer care was shut down across my health care system and I was deployed to work in any and all capacities in the emergency room of Mount Sinai in Brooklyn. At the peak of the pandemic, patients died in hospital hallways, barely receiving actual medical care—and they were far more likely to be patients without a doctor to advocate for them, an ability to navigate the health care system, or insurance.

A large part of disadvantaged survival from COVID was related to underlying health conditions, which can be directly related to poorer access to solid and reliable antecedent health care: uncontrolled diabetes, high blood pressure, heart disease, and weight were all factors that increased the risk of dying. The virus also spread more rapidly among low-income communities with denser populations: multigenerational dwellings, housing projects, and other living situations that involve higher levels of exposure to larger volumes of other people, as was typical of the neighborhood where I was deployed.

Besides basic access to care, hospital and death rates for COVID reflected racial disparities as well. Compared to White people who contracted COVID, Black and Hispanic people were almost twice as likely to be hospitalized and more than twice as likely to die. Native Americans were more than twice as likely to be hospitalized and three times more likely to die. We saw those statistics play out in our hospitals and on the front lines across the country. While the pandemic may have laid our country's health care disparities bare, they existed long before 2020. For those of us practicing medicine in the trenches every day, the issue of how to rectify health care disparities comes up on almost a daily basis. While I certainly don't have the big-picture answers, I start with the individuals in my world. My colleagues and I believe that we have the responsibility to do everything we can to prevent anyone from slipping through the cracks on the complex journey through breast cancer treatment. We pay attention to the factors that we can influence, where we can effect change, like access to quality care. In addition to system disparities, there are also specific biological factors that can affect breast cancer outcomes that, although we may not be able to change, we have to recognize and raise awareness about.

Breast cancer disproportionately affects Black women more aggressively, and they tend to have a worse outcome, on average, than their Caucasian counterparts. For years, the disparity has been attributed primarily to poorer access to expeditious and top-tier care. I have personally observed this, and our mission at Mount Sinai is to rectify this disparity within our own health care system, a system that is dedicated to taking care of one of the most diverse patient populations in the country, here in New York City.

For example, I initiated a research study investigating the length of time it takes women who present to our center with a new diagnosis of breast cancer to start treatment or have surgery, whichever comes first. We compared length of time between White women and Black women, in addition to comparing that time across women of different socioeconomic groupings and insurance types. We found that even within our center, where everyone has the same access to care, Black women and those with Medicaid or no insurance waited significantly longer to start treatment. I was despondent. How could that be in a center that treats all women the same regardless of ethnicity, background, or ability to pay? Clearly, the data did not reflect how we set out to provide care.

The apparent disparity led us to dig deeper. What we found is that when women are diagnosed with breast cancer, there are different reasons why the start of treatment or surgery may not be immediate. We often do additional imaging studies to make sure no other areas of cancer are present. We do MRIs in many, PET scans in some, and genetic testing in others, or some combination of tests that can guide us toward the best treatment pathway. Not all women need all these tests, but some do. Here's where the disparity comes in. Some women walk through our door having been shepherded by a primary care provider who has spearheaded their care and helped them get all of these tests done *before* even seeing us. By the time they show up in our offices, their case is delivered to us in a gift-wrapped box, tied up with a bow. Others come in with the bare minimum: a report on a piece of paper that shows a diagnosis of breast cancer. Or they show up at the reception desk with a lump in their breast and instructions from their doctors to come to our center, sometimes without an appointment, and we take it from there. In that second group of patients, we still have a lot of work to do to get ready for surgery or to start treatment, whichever comes first. We need additional tests to determine our approach— and those can take days or even weeks to complete. Unsurprisingly, that second group has a disproportionate number of patients at a lower socioeconomic level, including those on Medicaid or with no insurance at all. In those cases, there *is* a delay so we can do these additional tests that were not done before, so care that is ultimately given *is* equal and at the highest level across the board.

Beyond the testing delays, there can be additional delays related to logistics. Breast cancer *never* comes at a convenient time for anyone. While some decisions to minimally delay care are related to work obligations or even happy milestones like trips they had planned or family celebrations, some are related to a lack of resources or support that a woman might need to move forward with her cancer care. What if you are a single parent working two jobs and have few, if any, childcare options? Well, it's harder to get tests done in the small windows of time you barely even have to begin with. Appointments get made, tests get scheduled then canceled and rescheduled when a babysitter doesn't show up or a patient has to work overtime or has no way to get to the hospital. As health care providers, we devote significant resources to making sure that after a cancer diagnosis, care does not get derailed for patients with fewer resources. In our center as well as other top-tier centers across the country, there are often social workers who come in to babysit children while mothers get biopsies. We've provided rides for patients who can't make it to their appointments on time if they travel by public transportation from work. Many offer free legal services for patients who are being threatened with eviction or job termination—and help women in abusive situations that they didn't have the resources to leave before their diagnosis but are made worse by this new challenge. It's worth it to go to a place that is committed to surmounting any and all logistic obstacles, whenever possible, to rectify delays to cancer care for all patients.

With all that we can do to level the playing field for all women who need cancer care, we can't control the second part of disparity: biology. Within approximately 300,000 women newly diagnosed with breast cancer each year, roughly 15 percent of the general population and 15 percent of White women will have triple-negative breast cancer, which can be more aggressive and challenging to treat. Black women who develop breast cancer have a 30 percent chance of it being triple negative. When we look at poorer outcomes in breast cancer survival rates, we can often point to *both* socioeconomic and tumor biology factors. The fact that Black women tend to develop this aggressive breast cancer at younger ages, even in the absence of genetic predisposition, puts them at further risk for worse outcomes. Black women tend to be diagnosed at later stages, in part related to less access to screening but also due to the aggressive nature of the disease. The United States Prevention Services Task Force finally acknowledged the fact of this disparity in 2023, leading to their reverting back to recommending mammograms beginning at age forty to address it. The USPSTF didn't go far enough, in that they recommended the interval to get a mammogram to be every two years. All of our data points to the fact that yearly mammography, not every other year, gives us the best chance of picking up cancers earlier. A two-year interval makes no sense when trying to specifically target aggressive breast cancer that grows rapidly in young women. Too much can happen in that longer interval.

This confluence of factors—access, comorbidities, and biology—has led to the disparities in survival rates we see when we look at Black women with breast cancer versus their White peers. Current data indicate that mortality rates among Black women are 40 percent higher than those for White women, a sobering and disturbing end result of these disparities. The responsibility—*our* responsibility—is to address the factors that we can across our health care systems, so we can level the playing field of breast cancer care and work toward equal outcomes while using research to address biological factors as best we can.

34

MALE BREAST CANCER

EVAN MARGOLIN, JLL, vice chairman/tenant representation: Fortunately, my wife and I are still in love and she was rubbing my chest as we were lying in bed in July 2020, during the heart of COVID, when she said, "What's that?"

It felt like a grain of rice below my skin. Luckily, I had a dermatology appointment coming up. My doctor said, "I don't want to be an alarmist, but I don't like it." She made some calls and got me a needle biopsy appointment.

Soon after, I was sitting on a Zoom call and got two repeated calls from my doctor, so I picked up. She said, "I'm sorry to break this news, but you have breast cancer."

The way I looked at it was that lots of people have breast cancer. I certainly know some that lost the battle, but I know lots and lots of people who had a much better outcome. I didn't google. I didn't look at what a male mastectomy looked like or the scar I was going to have. I never looked at success rates after surgery.

I just thought, *I'm trusting my doctors, and I'm blessed to be in New York and have excellent access to care for this uncommon situation. I'm going to trust that*

I'm going to be okay until somebody tells me I'm not, and I can deal with that then.

Everybody's insecure in their own way. I'm not super insecure. I walk around with my shirt off. I have one nipple. I've had people tell me I can get a tattoo. If I get a tattoo on my chest, it's not going to be a nipple. I embrace it. The scar has sort of faded and it doesn't look terrible. If it starts a conversation with people, then I'm doing what I want to do, which is to spread awareness about male breast cancer. Tattooing a nipple or hiding it would defeat that purpose.

When it came to the medication to follow surgery, I didn't do much research in advance, so I didn't know what I was going to take. I didn't bother studying up on stuff that might happen down the road; I preferred to just wait.

When I found out that my Oncotype was low enough that I didn't need chemotherapy, we discussed tamoxifen. I was happy to take it, if it was going to help me to remain healthy. Then, I did a little research on tamoxifen for men and couldn't find much. It's kind of like my thoughts on TripAdvisor: people will only write something if they're not happy. The only things I could find were about men gaining weight and experiencing constant sweats and palm sweats and stuff like that.

When I picked up my tamoxifen from the pharmacy, for the first time in my life, I actually unfolded that little booklet with the tiny print. Everything was geared toward women. In any event, I unfolded it and started reading it. From what I recall, there wasn't any reference to men at all in the tamoxifen fine print booklet, which makes sense, as I understand that only 1 percent of breast cancer or less is male. I started taking it and thankfully didn't have any side effects.

Some men must know that male breast cancer exists, but I imagine there is also a decent percentage of them who don't. That's why I tell my story to almost anyone who will listen. I often find that men will be rubbing their chest while I tell them that my wife felt that little rice grain. I know that if they ever do feel something, they'll probably remember the story I told them. I always say that I love what I do, and I have a great job and achieved some great success, but I'm definitely not saving lives. This is my opportunity to save lives, and I take it seriously.

LAWRENCE D., senior private equity professional: I walked around with this bump for a long time. I google searched *male breast cancer*, and it said that typically it presents in and around the nipple. Mine was five centimeters from the nipple. It was smooth and small. Since I'd had cysts before and it presented like one, I thought it was just a skin cyst. I put some hot compresses on it and watched it. It didn't change.

I showed it to my skin doctor and my GP. They each saw it twice, and neither of them reacted to it. My skin doctor thought it was a dermatofibroma and wasn't worried about it, but since I pushed a little, he said, "Let's just take a biopsy." Two weeks later, I found out on MyChart that it was adenocarcinoma: breast cancer. Thankfully, it seemed to be stage I.

The first thing I did was google, and it looked like I wasn't dying at that moment. I contacted my sister-in-law, who is a gynecologist, and she recommended Dr. Port.

Soon after, I was on the operating table. I'm still here, and everything is quiet. Knock on wood it stays that way.

The thing I did not appreciate was that I was at the point in life when it's time to start chasing things down. If I had been thirty-five or forty, I would have understood the *It's probably nothing* mentality. I should have chased it down. That's the bottom line. I should have pushed people a little harder. "Hey, can we really take a look at this? This is making me nervous." It could have been addressed much earlier in its evolution. I got lucky, but that's a lesson I take away.

The way I wrapped my head around surgery was accepting that effectively I was losing a nipple. Unlike women, for most guys, they generally look the same. Even without the looking different element, you still need to wrap your head around these experiences.

The physical healing was not the hard part for me. The mental piece was. There are only 2,500 to 3,000 men a year that have this problem, so it's not like there's a statistical database of hundreds of thousands of people. I tried to keep things compartmentalized because it's really easy to let your imagination go nuts. It's not like you can go talk to all your girlfriends who can tell you about their experiences and how they got through it and survived. There's no one to talk to, and there's nothing to really read.

I had to learn how to rely on what I felt. If you feel great, that's half the battle in life, and everything feels okay. I looked at the statistical analysis of the exact sciences, and that gave me a lot of comfort. I got my Oncotype test back. Mine showed, statistically, that if I take tamoxifen, I have a 96 percent chance of having no recurrence. I just have to have faith in that number and hope that I'm not part of that 4 percent. I also have hope just thinking about all the evolutions and developments that will happen over the next ten years.

DR. PORT: Every October—Breast Cancer Awareness Month—women get bombarded with messaging to get their mammograms or get genetic testing or find out their risk. For women who have already been diagnosed with or survived breast cancer, those messages can bring back floods of memories from a profoundly difficult time. It can be overwhelming! But if one woman sits up and takes notice, if one woman does go for a test that then leads to early detection and cure, I'd say it's worth it. After all, breast cancer *is* the most common cancer to affect women. From my perspective, spreading the word about a disease that can affect half of our population makes it very relevant. You know who doesn't usually internalize the messaging? The other half of the population. Men, of course, might want to help raise awareness for the women in their lives. When football players wear pink cleats, that's who they are usually honoring and supporting: their wives, their sisters, their mothers, but not necessarily themselves. To be honest, that makes sense given the rarity of male breast cancer. It is this lack of awareness, as Evan and Lawrence said, that makes it all the more important to talk about it. To say it loud and clear for those who still don't know: men can get breast cancer.

Of the estimated 300,000 new breast cancer diagnoses in the United States each year, approximately 1 percent will be in men. That adds up to approximately 3,000 new cases annually, making it one of the rarest cancers that men can get and comprising less than 1 percent of all the new cancers diagnosed in men. If you ask a woman over forty to count the women she personally knows who have been affected by breast cancer, she usually has to go to at least two hands. If you ask a man the number of men they know, the answer is usually a blank stare and a response that they didn't know men could get breast cancer. This lack of awareness leads to what we commonly see in our male breast cancer cases: delays in diagnosis, later stages at diagnosis, and potential need for more aggressive therapies, related to those factors.

There are some well-known risk factors for male breast cancer that are also rare but important to know. Age is the first. The average age of a man getting breast cancer is older than for women; it's closer to seventy. Family history is also important. There is an increased risk of male breast cancer—approximately seven to ten percent—in men who have the BRCA2 gene, although, interestingly, there's no significantly increased risk with BRCA1. We do recommend genetic testing for all our male patients with breast cancer. That doesn't mean, however, that a male who gets breast cancer is likely to test positive for the gene. A man with breast cancer has about a 10 percent chance of testing positive, so, as with women, most male breast cancer is not attributable to a specific gene that we can test for. Other risk factors for male breast cancer involve syndromes and conditions that affect hormone balance, like liver disease that might affect breakdown of estrogen, or testicular disease that might lower male hormone production. There is also a rare condition called Klinefelter's syndrome where a male has an extra X chromosome (XXY, instead of just XY), leading to increased estrogen production. The risk theme here comes back to estrogen exposure, even for a man.

Male breast cancer should not be confused with gynecomastia, an enlargement of male breast tissue, either on one side or both. Gynecomastia is a condition that affects a large number of men, more commonly older men but younger men can be affected too. The main causes of gynecomastia are not the same at all as male breast cancers. In most cases, its development can be a side effect of prostate medication, some blood pressure medications, anabolic steroids, or certain ulcer medications. Obesity or smoking marijuana can also lead to gynecomastia, as well as many other factors. In young men or even adolescents, hormone imbalances can cause gynecomastia. Most importantly, gynecomastia, which is common, is not a risk factor leading to male breast cancer, which as we have already reviewed is extremely rare.

Because men don't get mammograms, male breast cancer is usually diagnosed based on physical examination. Herein lies the danger of delay: not knowing that men can get breast cancer usually leads to a dismissal of symptoms for a long period of time. A lump, a rash on the nipple, a mass in the armpit. I have seen all of these and more treated with creams, anti-inflammatories, and even antibiotics for months at a time. Other doctors have written them off as a skin condition, a pimple, an over-reactive lymph node—anything but breast cancer. The most common presentation of male breast cancer is a mass directly under the nipple, where the little breast duct tissue that men have lives. Because there is so little breast tissue, cancer can also affect the male nipple: inverting it, distorting it, or creating a rash-like appearance. One of my male patients, an avid surfer, thought for months that the nipple "irritation" came from surfing without a rash guard top and rubbing on the board. Because diagnoses of male breast cancer are often delayed, there is a higher likelihood of larger tumor size and lymph node involvement where a patient notices a "lump" under his arm.

DIAGNOSIS

Once we suspect there might be a case of male breast cancer, our immediate next steps are the same as for women. They usually involve a physical examination, a mammogram, a sonogram, and a biopsy of suspicious findings. Genetic testing should be done in almost all men with breast cancer. The majority of male breast cancers are ductal, and the majority, even more so than women, close to 90 percent, are hormone positive.

SURGERY

Surgery for men almost always involves a mastectomy, given that the majority of cancers are centrally located within the breast, and nipple preservation is almost never an option. Lymph node management parallels that for women; the majority of men get sentinel node biopsy or axillary dissection if nodes are already proven positive. Men almost never get breast reconstruction, but a few of my patients have had nipple tattoos. Many men with hairy chests don't feel that their incisions are readily visible at all and are quite comfortable with their appearances, even on a beach day with their shirts off.

In my opinion, patients with male breast cancer have even more need to seek out a specialized center and a surgeon who has experience with this disease. In 2001, my colleagues and I published the first series on sentinel node biopsy in sixteen male breast cancer patients. In 2008, we followed up with the largest series to date at that time: seventy-eight male patients. As a result of these two publications, especially the first one, which utilized a technique that was not only new but extremely new in men, I received a lot of male breast cancer patients who knew that my colleagues and I knew as much as anyone possibly could about their cancer and how to manage it—and, most importantly, had experience actually operating on their disease. I can tell you from having operated on dozens of male patients over the years, the surgery is different. Without getting graphic, digging into the lymph nodes is more challenging in a man, based on the different musculature and anatomy. The bottom line for men who have breast cancer is that getting to a medical center—and specifically a cancer center—where people are experienced with surgery for this rare disease is critically important.

TREATMENT

While male breast cancer does often present at a more advanced and later stage due to delay in diagnosis, there is every reason for men to be optimistic about their prognoses as well. Because male breast cancer patients are such a small group, they have been historically excluded from many clinical trials, so the actual data we have for male breast cancer are quite limited, which is common with rarer diseases. When it comes to treatment pathways for male breast cancer, we mirror the care of our female patients and extrapolate almost everything from there. Historically, that has worked very well in the process of making decisions about Oncotype testing, chemotherapy, antihormonal therapy, and even the possible need for radiation after a mastectomy. There have been some data showing that male patients, especially older ones, have a lower compliance rate with much-needed antihormonal therapy compared to women. Men can experience the same side effects, such as hot flashes, sweats, and decrease in libido.

For our male patients, the breast cancer experience is different. It should be. In our center, we have a subset of equally soft changing gowns that come in brown instead of the usual pink. We have private sections of our waiting areas for patients already in their gowns, so neither men nor women have to feel uncomfortable partially clothed in the presence of the opposite sex. Men and women typically handle the diagnosis differently. Some men are bashful and ashamed to talk about a cancer that they thought only women get. Others are open and hope to pay it forward by raising awareness like Evan and Lawrence. Over the last twenty-five years, I have taken care of a husband and wife, both with breast cancer, five years apart, as well as a brother and a sister. Everyone handles their own diagnoses differently, as one would expect. The most important thing that we can do is support our patients and destigmatize the disease. While it's a cancer that we see far more predominantly in women, it's not only a women's cancer.

A NOTE ON TRANS PATIENTS

mportantly, trans men are at risk for getting breast cancer, even if they have had top surgery. Their risk is less with most of the tissue removed, but higher than that of a cisgender male. Removing breast tissue for top surgery is not performed in the same way that a cancer-curing mastectomy is. Much breast tissue can be left behind, leaving some risk. While the last thing a trans man might want to do is a mammogram, screening and awareness of the possibility of getting breast cancer remain important. Similarly, trans women can also be at risk, especially because transition usually involves taking feminizing hormones. Breast enlargement from these hormones is common, and screening should be considered as well.

35

TALKING TO OUR CHILDREN, FRIENDS, AND COLLEAGUES

WHAT DO THEY NEED TO KNOW ABOUT YOUR BREAST CANCER?

Two friends, Gina and Michelle, were diagnosed and treated for breast cancer at the same time. Here they share their experience of how they spoke to their children, of different age groups, about their breast cancer diagnoses.

GINA RUNFOLA, director of operations: My kids were twelve, fourteen, and sixteen. As most people know from my personality, I'm more direct. There is no feeling sorry for myself. I'm a *Shit happens and we've got to pick ourselves up* type of mom. I'm not one to shield my kids. Life is life. Life is messy. I think when you expose them, it builds resilience. So, I basically told the three of them. I pulled the three of them together, and it was very quick. I told them right after I found out because our community is very small. Where they go to school, we have a lot of friends who are moms of their friends. The last thing I wanted was for them to find out overhearing someone else talking about it. I have a friend whose daughter found out that she had ovarian cancer overhearing another mother in school. There are also people who suffer in silence, and I didn't want my kids to feel that if they are experiencing hardship, they have to be ashamed by it. I remember saying to them that Mom has breast cancer. I reminded them that there are like fifty people that have had it in our community, and they're all doing well. I gave examples of people. To be honest, I was more worried about telling my parents than I was my children. You can understand that as a mother, the kids are resilient.

I said, "There are three things I want to tell you: Don't say 'Why me?' It's 'Why *not* me?' Because I'm not better than the person next door. Don't be angry. And don't feel sorry for me, I, we, will get through this." I said, "You're going to see me sick. You're going to see ups and downs, you're going to see a different side of me, but I'm going to be okay, and I'm going to get to the other side of this." And then I said, "Do you have any questions?" And they were like, "No, we are okay too, Mom."

The best thing you can do is show your children that you can either get bitter or better from a bad situation, and it's how you respond to it. I couldn't control what happened to me, but I can control how I responded to it.

MICHELE S., happily retired grandmother of two: I have two girls, twenty-four and twenty-nine. They were much older than Gina's children, basically adults, but I'm still their mom. Right before I was diagnosed, I had just visited two friends from high school. I felt bad because they had hard lives. I remember that day, I was walking the dog, and I said, "Thank you, dear God, I have such healthy children, and we are all okay, and our lives are so good. I'm so blessed." The next day was the mammography, and everything hit the fan.

The first couple of days, I didn't say anything, just because, unfortunately, my daughter's husband's young cousin had just passed away from breast cancer at twenty-nine years old. I knew that I was going to freak my daughter out. I just sat them down, and they were looking at me. I said, "Listen, I just want to tell you, I'm going to be absolutely fine." Right away I said that because my daughter went to the wake and the funeral of that young girl, so I wanted to start with that. I told them, "I just have to get some treatment and radiation, and I am going to be fine, it's nothing."

I just kept telling them, "I'm going to be fine, I'm going to be fine." I know they were scared. The younger one acted like nothing was wrong with me. People would say, "How's your mom?"

"Oh, she's fine. She's absolutely fine."

Just repeating what I said, but I could tell it bothered them. I just kept trying to reassure them, even when I would fall asleep on the couch after chemo. Usually in the afternoon, I would just conk out, but as soon as I heard the door open, as soon as I heard my daughter come in, I would get up. I also never walked around without my wig. I didn't want them to see my bald head. I just didn't want them to see me that way.

I waited longer to tell my parents because I said, "Oh

my God, my mother and father are not going to be able to handle this." As hard as it is to tell your children, it's harder to tell your parents. It's just not the right order of things, and it's hard for a parent to watch their child go through something, no matter the age. Even my father said to me, "Don't come here without your wig on. I can't take it to see you that way." I felt like it was important for me to protect everyone else in my family, even though my kids were older and, of course, so were my parents. For sure it was hard for me, trying to protect everyone else, but that has always been my way.

DR. PORT: When a patient with breast cancer is also a mother—which many of my patients are—she often goes right to worrying how her diagnosis will impact her children, from the logistic to the existential. Everything from *Who will take my children to school while I am in the hospital after surgery?* to *Will I be around to watch my children grow up?* Thankfully, I can provide almost all of my patients with the reassurance that yes, if we do everything right, breast cancer will not keep them from being with their children for decades to come. As parents diagnosed with breast cancer, your job is often to transmit this optimism and reassurance to your children. Timing is everything. One of the earliest decisions you need to make about anything as a parent is what to share with your children and when.

Kids are smart. Most kids today have access to a lot of information beginning at a very young age. According to Common Sense Media, about half of American children now have cell phones by the age of ten, and over 90 percent have one by the age of fourteen. Our kids are able to read things online. They google things. They're on social media. They also hear things. Critically, they're getting all of this information without the knowledge or supervision of a parent. Twenty-five years ago, when I was starting out, the internet and social media were barely around, and most children couldn't access information about breast cancer beyond what their parents might communicate. Many more people kept the diagnosis from children of all ages. In those days, I had mothers who would say something like, "Dr. Port, I would like to put my kids on the school bus in the morning, come in for my lumpectomy surgery, and be home in time to pick them up without them knowing anything happened to me." That sort of thing was more doable then, since the potential avenues for children being exposed to information were limited. Now, it's more challenging, unless you are committed to telling very few people, ensuring those people tell no one, and being vigilant about your children having no chance of overhearing voicemails, reading emails, or even seeing your search history on the family desktop. In the age of information, it's much better to provide the information directly and craft the narrative yourself. Conversations should always be age appropriate, but they should all lead with well-founded optimism and a positive outlook. Then, hopefully, your children will take their cues from you.

There are common themes to take into account for children of all ages. There are also different ways to share information with different age groups.

Most children of all ages want reassurance. They want to know you'll be okay, that you'll be there for them. They also want honesty, especially if they're older. The issue is that in the most extreme cases, reassurance and honesty can sometimes conflict. It's important to strike a balance between them. One of the main sources of reassurance—for you and your children—is having a plan of action and a path forward. It's not reassuring for children to hear about a cancer diagnosis when there is no plan. When you're still formulating the plan yourself, you may choose not to tell your children until you have a clearer idea of what the future will look like. Here's the thing: if you don't want your children to know right away, my main piece of advice is to keep your circle very, very tight. If you want to keep your diagnosis a secret from your kids, do not tell people who don't need to know. Anyone. Because, at some point, even if you have sworn that person to secrecy, that person might have a conversation with their husband or wife or significant other or partner that gets overheard by their child who happens to be in your child's class in school. The worst scenario is for a child to hear about your cancer diagnosis from someone else. For children, learning that secrets have been kept from them creates mistrust and can lead to future fearfulness and anxiety that they're not being told the truth. Once you have a plan, share it with your children immediately, so it comes from you.

Lead with age-appropriate honesty and optimism. With younger children, be careful with the word *cancer* because it can have so many negative connotations. A child may hear the word and know someone who has a family member who died from breast or some other cancer and struggle to distinguish between the two cases. Most children will take their cues from you. If you convey fear and pessimism, they will pick up on that. If you are anxious, they will be anxious. I've had many mothers come in newly diagnosed with breast cancer who've been on multiple medications themselves for anxiety or depression for years tell me, "I am afraid to tell my child about my cancer as they are already very anxious by nature." Talking to your children about your diagnosis is when the number one job of being a parent, protecting your children, should really kick in. In some cases, when the parents really feel that, for whatever reason, their child is struggling with the news, I have offered to talk to the child on the phone or even have the parents bring them to my office, so I can convey the reassurance that an anxious parent may not have been able to provide. Sometimes hearing from me or seeing for themselves what treatment entails can be grounding for a child, but other children might not be up for that level of exposure. Your relationship with each of your children may also be different, depending on their ages and their individual dispositions. It's still important to make sure that if one child knows about your breast cancer diagnosis to any degree, they all should know to prevent strains in the siblings' relationships with one another.

I should also say that having your plan doesn't mean you have to know the *whole* plan. You may just know the first concrete steps. Many women have surgery not knowing if they need chemotherapy or radiation to follow. Women who are having chemotherapy first may not know what kind of surgery they will ultimately need. Most children, especially young ones, will not be able to absorb or retain tons of information about long-term plans and possibilities. My advice is to take a step-by-step approach. Of course, an older child might want more color and details. So here are some general age-related guidelines I've given my patients over the years—and, more importantly, the subsequent feedback they've given me.

FIVE YEARS OLD AND UNDER

Most children under the age of five can barely understand the concept of cancer, even in the simplest terms. They also, in most cases, have no access to outside information. They want to know what your situation will mean for *them*, only in terms of the immediate future. You might tell them, "Mommy will be having surgery and will stay in the hospital for one or two nights. Grandma and Grandpa will take care of you, and then I will come home, and you will see that I will be fine." Going into statements like, "I am going to have surgery, and I might need treatment after, and I might look funny and lose my hair . . ." are far above and beyond what a child wants and needs to know. I have had patients tell me their child is exceptionally smart and so mature. While I assure them that I'm sure their children are special, I stress that no five-year-old will remember what might happen a month from now. Once you have your plan, talk to your child in simple and concrete terms. Focus on the immediate and near future. If you are having surgery first, you can also explain the things you will and will not be able to do after surgery. You might reassure them that you'll still be able to put them to bed and read stories. You might say that you'll be able to pick them up from school soon after. With a bigger surgery, you might say, "We can't have our pillow fights for a little while, and I won't be able to pick you up right away, but I will be able to do that again soon!" Concrete examples are enough to provide the comfort and reassurance that most children in this age group need.

AGES FIVE TO ELEVEN

This is the group that now has more access to information than ever before. Many children on the older end have phones that allow them to communicate in ways most parents can't control.

They can read, and they have a greater capacity to understand time and remember what they're told. Smaller children will often ask, "Will it hurt?", "Will you be able to play with me?", or "Will you look different?" But as kids get closer to eleven, many of them are googling questions on their own. For the most part, this age group will still want to know how your diagnosis will affect them. They'll also be most vulnerable to picking up shreds of information from overheard phone calls, voice messages from a doctor's office that you accidentally play back on speaker phone, and inadvertent feedback from friends and family. You want to get ahead of all that by telling them as soon as possible, before they find out some other way. Again, leading with reassurance and optimism is the best way to go.

AGES TWELVE TO SEVENTEEN

Children in this age group, especially on the older side, tend to have complete capacity to understand your diagnosis, but the method and timing of the delivery are still just as important. They still want to know how your cancer will affect them. How will it impact their social lives, extracurriculars, and school responsibilities? In our current day and age, teenagers are extremely tech savvy. What do you say when your child has googled everything about breast cancer and has a million questions? Again, the way to keep everyone from getting lost down online rabbit holes is to provide reliable, empowering information. The goal should be informed optimism. Teenagers will also have capacity to absorb information about longer term treatment plans and can understand more details about additional therapies that you may or may not be need in the near future.

The biggest danger for this age group and the next is what we call "parentification": a role reversal that turns a child into a caregiver, which can be extremely stressful for the child. Many women do have a limited support system. You might be a single mother, or you may not want to share your experience with others, or family may be far away. But think about who else you might be able to call on to support you through appointments and help you make your decisions. I would recommend thinking twice before bringing a child under eighteen to appointments as your support system. It's a lot of pressure for them. It can also make it challenging for us, your doctors, to speak frankly about potential negative outcomes in their presence. We have to inform our patients about the rare but real risks that surgery involves, the side effects of treatment, and other negative potential long-term events, but that information can be extremely stressful for a child to hear.

Daughters may have a particular sensitivity to your diagnosis and make the connection to what it might mean for them. Be mindful and be prepared for questions in this arena. If you have undergone genetic testing and are positive, know that it could have even more significant implications for a child, especially a daughter. My opinion is teenagers are too young to handle this information and be burdened by the fact that at some point in the distant future, they should probably be tested themselves. I always say to let them be kids, but every family is different. I had a patient with breast cancer who was BRCA positive and so riddled with guilt about what she might have handed down to her daughter that she had her tested at seventeen years old, praying for a negative result to ease her own anxiety. When the results came back positive, she was devastated and brought her daughter in to see me. It was clear that this teenager, not even a senior in high school, was completely overwhelmed and unequipped to deal with the implications of BRCA-positivity. She should have been dealing with college applications, but all of a sudden, she was thinking about some future need to have a mastectomy. In this situation, my goal was twofold: to reassure the mother that this was not her fault and to reassure the daughter that, unless she ever felt something in her breasts that concerned her, she needed to put the results out of her mind until she turned twenty-five. Then, we would revisit them.

AGE EIGHTEEN AND OVER

An eighteen-year-old is legally an adult, but they may still be a child in many ways. All of the same concerns that a young child may have about their parent's well-being still apply, albeit with infinitely more understanding.

The discussion about how to tell adult children about a diagnosis often involves logistics, since children may already be off living on their own, a distance away. Is it best to break the news in person, or can it be done by phone? Should you wait until a child in college finishes midterms or final exams? What about a child who's deployed in the military and may not even be reachable?

Again, my advice is to always be up-front. Dialing up or dialing down the perceived risk and need for the child to "come home" is very variable, and very dependent on the family dynamic. Always strive to maintain some sense of normalcy for your children. Do all three grown children really need to take off work and support the other parent on the day of surgery or can they stay at work, take their tests, and focus on their own lives with appropriate notifications and visits as needed? It's hard to say. Of course, I have seen both work. I've also seen both not seem to work, where the children end up bearing an outsize brunt of stress and responsibility for caregiving.

Ultimately, you know your children best. You need to use your judgment in communicating with them about your breast cancer.

TELLING FRIENDS AND FAMILY

The importance of love and support from one's friends and family when going through a difficult time, such as a breast cancer diagnosis, is immeasurable. In today's world, the way people tell others that they have breast cancer can range from something as intimate and personal as a phone call or an in person visit, to publicly announcing it through an Instagram post. Your first decisions related to sharing should be regarding how private (or public) you want to be. Do you want everyone to know? Or just a few people from your inner circle? For sure, part of the decision-making regarding the desire or need to share might be dictated by the nature of what your journey through breast cancer might look like. It is genuinely difficult, but not impossible, to keep a breast cancer diagnosis a secret when chemotherapy is involved because the treatment goes on for months and sessions usually involves hours of time. And then there is the associated hair loss. The cold cap certainly can help preserve privacy by minimizing change in appearance. But if your situation does not require chemotherapy, it's completely feasible to get through a breast cancer diagnosis and treatment with very few people knowing about it if there is, for example, just a small surgery with minimal recovery and downtime, and a pill to take afterward.

The choice is of course yours to make. But remember, telling one person can often involve telling all. Recently, a patient I'll call Jane came into my office newly diagnosed with breast cancer and said, "I know we know a lot of people in common, I don't plan on telling anyone, and I just wanted to let you know that." I reassured her that, of course, I had no intention of speaking to anyone else about her diagnosis much less divulge that I was treating her. And that HIPAA laws absolutely forbade me from doing so. About a week later, I was at a dinner with some of these mutual friends, and as I walked through the door, one approached me and said, "I know you are taking care of Jane, how is she doing?" I was caught off guard. Perhaps she changed her mind? Was this a test? I just smiled and changed the subject. When I saw Jane the next time, I mentioned to her that this mutual friend had approached me, but reassured her that I said nothing and gave her no information from my end. "Yeah," she responded. "I did end up telling a few people. And I guess they each told a few people. It's okay," she went on. "It was just too hard to keep it under wraps. And I also didn't want my kids to feel like they were sworn to secrecy since it's their story to tell, too, that their mother has breast cancer."

Many people choose not to keep their diagnosis entirely secret feeling that perhaps they can help others who might go through it in the future. This is not exactly public, announcing it to the world, but not exactly private either. And this is a path that many patients choose to take the pressure off close friends and family members regarding accidentally divulging that their loved one has breast cancer. And to pay the support forward. Whether you decide to keep the circle tight or share widely, both have their benefits and downsides. Keeping a tighter circle, telling your partner and one friend or so, might make it easier to guard privacy. But then there are fewer people to lean on and support you through your process, whatever that may be. Sharing widely allows for more support, but there are also more people to keep informed, who might constantly ask for updates and potentially transmit some of your information to an even wider circle than you intended.

Just remember that together with those closest to you, sharing your diagnosis is your decision to make.

TELLING WORK

Telling your colleagues at work is another decision to make, and again, might be, in part, dictated by what your journey involves. Having surgery or treatment may require time off. One or a few days' absence from work does not usually require a discussion with work colleagues or supervisors and is often easy to work around if you don't want to reveal your diagnosis. However, an operation like a bilateral mastectomy, which requires overnight stay in the hospital and often weeks of recovery, is a different story. Most of my patients who work require three to four weeks off to recover from these operations. Getting back to work after a more extensive surgery can involve taking that full period completely off, especially if work involves a manual component. I have seen others back on their work emails and Zoom calls while still in the hospital, which I don't recommend! As I told Laura back in Chapter 18, "Don't text on drugs!" And most women do need pain medication following larger operations. But in our world where a lot of work can be done remotely, many do choose to join back into work activities in a limited way even a week or so following surgery, taking a Zoom call here and there, or sending out work emails.

Before you decide whether or not to tell work colleagues, my advice is to first get a sense of what your path through breast cancer treatment might look like, and then to take stock of what your workload requires, and whether or not continuing work in some capacity is feasible and desired. Having some idea of your treatment path can then guide you as it relates to discussing your diagnosis and your plan with those at work. If your treatment will involve a longer course with significant absences or you prefer to tell your work colleagues regardless, know that many of my patients have told me they were pleasantly surprised by the supportive and kind reactions as well as accommodations that were made for them. As you might recall from earlier chapters, Gina was able to work three days a week, recover from her chemotherapy over the weekend, and preserve structure in her life, which she very much desired. Stacey Griffith, whose work was highly physical in nature, needed to take months off and return when she was ready. Because breast cancer is so common, it's likely that the work colleagues you speak to, even the men, will have also been affected in some way: a coworker's sister, a boss's wife.

SILVER LININGS

BREAST CANCER CAN CREATE A TEACHABLE MOMENT

Many of my patients say that while they are "in it"—actually going through the experience of breast cancer—it is almost impossible to imagine that *anything* good could come from it. While no one wants to get a cancer diagnosis, over my many years in practice, I have heard from so many patients who tell me that their diagnosis did bring something positive into their lives. Take Gina and Michele, the two moms you heard from in the last chapter. They met the day before surgery, becoming each other's saving grace during a very difficult time.

MICHELE S., happily retired grandmother of two: We were waiting in the nuclear medicine department, for that injection you have before surgery, and I heard this voice. I actually thought it was somebody I knew from Staten Island who I would be eager to see because you feel lonely, you feel scared, and any familiar voice or face would be welcome. Of course, it wasn't the person I thought was there. This woman came over and ended up sitting right near me, and she was reading this book. I just knew that I wanted to talk to her.

GINA RUNFOLA, director of operations: Neither of us knew what the other was there for. It's the nuclear medicine department at Mount Sinai. There are thousands of patients there for different procedures every day. It's not just breast cancer, like you could have been there for something completely different.

MICHELE: So, then I think we started chatting a little bit, and I asked what you were there for—

GINA: No, Michele. You and Eddie, you came and sat right next to me. I can't explain it, but her husband was dressed nice, and my husband was dressed nice, and we had similar appearances. And she comes and sits right next to me. There's forty seats in the room, and she comes and sits right next to me. I'm thinking, *She's either going to be really friendly or straight-up weird.*

MICHELE: Then I asked her where she was from. She told me Staten Island, which was so strange, since that's where my friend whose voice I thought I heard was from, and who ended up being Gina. Then I asked, "Who's your doctor? What are you going for?" It started from there, right?

GINA: I started crying. I said, "I just got diagnosed with breast cancer." "Who are you seeing?" "I'm seeing Dr. Port." When she told me she was using her, too, I'm thinking there's safety in numbers. We were trying to get in as much conversation as we could because I was getting called in first. We quickly established that our surgeries were on the same day, the next day. As I got called in, Michele said, "Can I have your number?" So, we exchanged numbers, and then off I went.

MICHELE: On the next day, our surgery day, she was the surgery right before me. That morning, when we came in, she was with her husband and her sister. I was just with my husband. It was like, "Okay, good luck! Good luck." I didn't want her to go into surgery yet. I needed to spend more time with her, like to reassure me this is gonna be okay, we're gonna be okay.

GINA: After surgery, we texted, and we spoke on the phone comparing the diagnosis, comparing our fears, crying, you know, just dealing with the unknown. There's a lot of emotion, as you can imagine. There's a lot that comes along with the diagnosis, right? Michele was the only person who knew in that moment how I felt. I think, Michele, the same was true for you.

Even though we were worried about ourselves, we also wanted the best for the other. For our postop appointments, we were also scheduled the same day. We went in together, and I went in to see Dr. Port first. She told me I needed more surgery—the margins weren't clean—and I chose a double mastectomy. I came out, and Michele and Eddie were sitting in the waiting room, and I told her. She is sitting there in the waiting room about to go in next. I said, "Michele, I don't have the best news. I didn't have clean margins." And she started to cry, and she hugged me, and she was so sincerely upset for me. It was just the most sincere hug. I can't explain it, but she felt my pain. Then, I was happy for her that she had clear margins and was able to move to the next step.

MICHELE: Then our chemos are the same days. I had my appointments on Mondays, and she scheduled her appointments for Mondays. Back-to-back chemos in the room, sometimes next to each other. My husband and her husband hanging out and the two of us hooked up to these ports. It was basically like a chemo cocktail party!

GINA: We wanted to kill ourselves, but it's funny when we think about it.

MICHELE: I think both of us were getting treatment on Christmas Eve, an important holiday for us. We helped and supported each other in ways that other people couldn't or didn't understand. Some people say the stupidest things. I love women but some of them are really dumb. Like this one friend asked me, "Why are you getting treatment so long?" And another—and I swear this is true—called me and said, "I'm sitting by the window, and my mother's friend's daughter is around my age, and she lives next door to me, and she had breast cancer, and they are taking her out in a body bag."

GINA: Then, of course, my phone rings, and it's Michele telling me this. I mean this is an extreme example, but people say such inappropriate things.

MICHELE: I was in Wegmans, and I'm getting fruit. One of my neighbors said, "I never knew somebody that had to go to treatment as long as you did." *Okay, thank you.* I put the fruit down.

GINA: We started weeding people out. You know not to call her, and you know not to call that other one because certain people don't make you feel good. So, find your Michele. Find the person that takes you out of that and makes you feel better and knows.

DR. PORT: That day in the pre-op area, a bond formed. Michele and Gina have spoken almost every day since for seven years. They have attended all of each other's major milestones: birthdays, anniversaries, family weddings. Michele's husband gave her a surprise sixtieth birthday and, besides family, Gina was the only friend invited. He actually asked her to make the toast at the dinner. Now, they schedule their follow-up appointments with me back-to-back, so they can come into the city together and make a day of it: exam, mammogram for Michele, who had a lumpectomy, shopping, dinner.

As both have told me, together and separately, they are blessed to have each other in their lives, and if that could only happen through breast cancer, so be it. They thank *me* for giving them the gift of each other, but I had nothing to do with it. Both of them were open to finding a silver lining through their diagnosis. Maybe it was a coincidence or maybe it was meant to be, but they have become each other's greatest cheerleaders and confidants, more like sisters than friends.

As I've said before, my patients inspire me. . . . Here are some more pieces of advice and the incredible experiences that they have chosen to share. The profound, teachable moments that I have then been able to pass on to others.

SARRAH STRIMEL BENTLEY: JOY MINING

You say silver lining, I call it joy mining. I think back through my breast cancer journey and how I was falling in love at that time. I was bald and farting from chemo, but I was falling in love. Every morning, my then-boyfriend would come to romance me. He proposed to me bald, during chemo. I was having a romance story, and then I was having the experience of preparing to be a mother.

I say mine the joy when you're in the shit. You're going through this, but there is joy everywhere. Sometimes you just have to look harder when it's covered in you-know-what.

Of course, there were days that I sobbed on the floor in a ball because I was in pain, but there were also moments that were incredibly celebratory and joyful that filled me. Those were the moments that kept me going. I'll never forget when Dr. Port said, "It's a marathon, not a sprint." I do not run unless I'm chased, but I know marathon runners have cheerleaders along the sidelines, and that gets them through the race. I found the joy in my people. The joy could be anything that happens in your day because that's going to get you to the end of the marathon. Find your signs. You're amazing. Drink your Gatorade and remind yourself you can do it!

CHIQUITA CARTER: KEEP FIGHTING

When you have cancer, some people see it as a death sentence. I would say that you have a lot to live for. For me it was my son. Keep fighting because the outcome could be better than you expect. There's a chance for you, so don't give up. If I wouldn't have fought the first time, I wouldn't have gotten married. I went through a lot, but life is good for me now. Since getting better, I joined the breast cancer Sisters Network in Tampa Bay. We're a network of Black breast cancer survivors. We get out in the community, and we put the word out. There are other things I want to do like drive people who can't to and from chemo. I wrote and published a children's book, *I Feel Blue*, and I had always wanted to do that. I'm excited about life now. I wasn't really living before. I was living to work. I didn't really appreciate my birthdays. Now I celebrate them because I get to live to see another year.

MARY JOE FERNÁNDEZ: DON'T SWEAT THE SMALL STUFF

At the beginning of the journey, you're more susceptible, more on edge, but eventually, you settle back into the real world. Back to living, back to focusing on my job, children, my husband. One thing I really learned through this is that I'm the one who went through it. As much as I had the best support team in the world, it was different for them, and my perspective changed. I try not to sweat the small stuff anymore, but I can't expect everyone around me to not sweat the small stuff to me. When my husband has a rough day at work, I'm like, *What's the big deal?* But it is a big deal to him. I've learned to see it, as he didn't go through the same thing exactly. I'm okay with that now. My perspective is different now. At the end of the day, health is the most important thing.

KATHERINE ESKOVITZ: GRATITUDE AND THE CLUB SHE NEVER WANTED TO JOIN

The first thing Dr. Port told me when we first met was: "You are going to be okay." I held on to that when smaller fears crept in—claustrophobia during an MRI, pain from a nuclear injection or insertion of a surgical wire, post-op nausea. Humor helped. Slowing down helped. Gratitude helped. Gratitude surprised me most. In my vulnerable state, I noticed every kindness: a stranger lifting my suitcase; a salesperson asking about my day; a nurse saying I looked too young for my birth date. After my diagnosis, these moments seemed to multiply. My perspective shifted in a good way. A friend told me her MRI was unexpectedly peaceful, even moving. Her mindset reframed the test for me, making it less frightening—and more meaningful. Later, I shared her words with another friend facing the same procedure, and they helped her too. A positive and hopeful mindset, visualization exercises, and meditation truly helped to transform unpleasant experiences. Sometimes I felt afraid and pretended to be strong. Even faking it made me stronger. I'm still in the middle of my cancer journey, but I already appreciate these silver linings. So many brave women I admire faced this before me. I feel a part of a club I didn't want to join, and yet I have experienced so much love and wisdom as a member.

LAWRENCE D.: GETTING HEALTHY

I've dropped fifteen pounds since my diagnosis and I don't think Dr. Port removed fifteen pounds of flesh. I never smoked, but since the surgery, I haven't even had a drink except for Passover. I eat healthier, I'm making sure I'm eating the vegetables and the fruits. I'm more physically active. I'm just taking better care of myself and not taking my health for granted. I feel like I'm twenty-five years old again and it's crazy. If I hadn't had this experience, I probably wouldn't have gone to this extreme. Once you clean yourself up, you don't miss it. The idea of eating meat, a filet? Not interested. We eat a fair amount of fish, low-mercury fish, like halibut or cod, and occasionally some chicken. My mom makes lamb stew sometimes and I'll eat that, but I haven't had a grilled piece of meat since my diagnosis. I can't promise you I'm going to live longer, but I hope to feel better in the process. And my advice, whether you have breast cancer or not, is to do the same. Do something to make yourself feel healthier.

VIVIANA FIGUEROA: FAITH

I believe my faith played a very big role along the way. I was born and raised strict Christian Pentecostal. As I grew up, I ventured out to go to nondenominational gatherings in New York City. I always just had a conscious faith that God was with me and had a close relationship with God. When I was diagnosed, it rocked my faith. I know that there's a plan and a purpose for everything, but when you're in the middle of it, in the thick of it, you start to have a lot of questions. I started to judge myself for having the questions. I was forty-one, but as I look back, I do see that there was a plan and a purpose. Even something as simple as this: I feel like God is opening doors for me to share my story on a much larger platform.

I got diagnosed through a virtual visit and I had to hop off the visit because I had a work call. I was a disaster. I was an emotional mess. I called my boss to say I couldn't make the two o'clock because I was just diagnosed with cancer. They told me I needed surgery, radiation, and chemo. I started to make phone calls to friends and family. I did want to go down the Google rabbit hole, but at the end of the day, I was washing some dishes in my kitchen and I felt God's presence impressed in my spirit. God showed me: "You're going to be fine." I had some sort of vision of myself telling my healing story in a stadium that was bigger than Yankee Stadium.

I believe that all these opportunities, like going on PIX 11, SurvivorNet, and other interviews, even participating in this book, are all opening doors and leading the path to be able to share that yes, there's medical research and yes, God uses the hands of doctors, but there's also His healing power.

DR. PORT: I am so happy, through this book, to be able to share these pearls of wisdom my patients have imparted to me over the last twenty-five years. My patients have reaffirmed my optimism by sharing their own. They have shown me time and time again that even something as terrible as a cancer diagnosis can have a silver lining.

ACKNOWLEDGMENTS

My sense is that for all books, it takes a village to bring them to completion: the writing, the editing, the publication. My village was all-in for this mission to provide the most updated information for women on breast health and breast cancer, and I owe them all so much gratitude and thanks.

First, thank you to all my patients who have imparted so much wisdom along the way—the wisdom that I now share here—and for giving me the honor of caring for you. Jill Martin, you started this. You brought me on TODAY to help tell your story, and it was a lightbulb moment for me: a new paradigm of combining patient experience with doctor expertise to impart information, optimism, and yes, the best breast advice.

Eve Claxton and Iman Fah Traore. Eve, we go back to working together on the first book ten years ago. You are an amazing editor, advisor, connector, and friend. As with all things writing, I could not have done it without you, coming through when I needed it most. You also connected me to Iman.

Iman, your brilliance and insights and your work on collating, recording, and editing the interviews, which are the heart and soul of this book, were indispensable. And while we worked together across an ocean, it always felt like we were sitting together chatting at the table. Thank you for your immense work and contribution.

Daryl Roth, my ultimate connector and lifelong friend, thank you for your support and for putting me in touch with Pam Friedman, a former agent and now also a close friend, who connected me to my agent, Erica Silverman. Erica, your insight and patience, along with your willingness to work around my crazy surgeon schedule in our calls and meetings, were so very much appreciated. You are a brilliant, thoughtful agent who I am so lucky to have in my corner. Daryl, Pam, Erica, you are likewise older sisters who believed in this work and made it happen. Rebecca Salzhauer, who also edited, thank you for all your insights and help, and her amazing mother, Amanda Salzhauer, a life coach, and my dear friend and always advisor. Marcy Engelman, you are the most incredible publicist and were instrumental from the beginning. I am so appreciative of your guidance and friendship.

HarperOne: Where do start? First, the incredible Judith Curr. Never could I imagine being in the hands of someone like you and your team. I am so grateful to you and feel that working with you and your team was meant to be the perfect fit. Gabi Page-Fort, my editor at HarperOne, you helped make this book the best that it could be through your insights, guidance, and brilliance. It has been such an honor to work with you. Thank you also to: Ryan Amato, for your help and, as always, Adobe Acrobat tech tutorials that I was in frequent need of in the editorial process; Paul Olsewski, an amazing publicist, and Aly Mostel and Ann Edwards from marketing; Stephen Brayda and Lindy Kasler as well as Sarah Kellogg who designed the beautiful cover; and Alison Bloomer, the interior designer who made the book's appearance exactly what I had hoped for. Also, thanks to Suzanne Quist, Cristina Nunez Blanco, Justine Gardner, Amanda Irle, Sierra Delk, Stephen Callahan, and Anna Brower.

To my family: my husband Jeff, kids Zack and Lauren, and my parents Loni and Jeff Rush. There are not enough words to thank you for your support and encouragement along the way. And for always doing the driving, Jeff, so I could write. And to my work family at Mount Sinai, my second home: Raina Caridi, Ali Sackett, Nicole Hall, Jane Wallace, Deborah Orringer, Courtney Sullivan, and Mark Taylor. Along with my amazing partners and colleagues. We are a team. And you have all played such a huge role in the care of our patients, together providing the highest level of breast care.

REFERENCES

CHAPTER 1

Hoxha I., Sadiku F., Hoxha L., et cal. "Breast Cancer and Lifestyle Factors." *Hematology/Oncology Clinics of North America* (2024), Volume 38, issue 1.

Jun S., Park H., Kim U. J., et al. "Cancer risk based on alcohol consumption levels: a comprehensive systematic review and meta-analysis." *Epidemiology and Health* (2023), Volume 45.

"What You Can Do to Meet Physical Activity Recommendations," Centers for Disease Control, updated December 4, 2025, https://www.cdc.gov /physical-activity-basics/guidelines/index.html.

CHAPTER 2

Morch L. S., Skovlund C. W., Hannaford P. C. et al. Contemporary hormonal contraception and the risk of breast cancer. *New England Journal of Medicine* (2017). https://pubmed.ncbi.nlm.nih.gov/29211679/.

Rossouw J. E., Anderson G. L. , Prentice R. L. et al. "Risks and benefits of estrogen plus progestin in healthy postmenopausal women: principal results from the Women's Health Initiative randomized controlled trial." *Jama* 2002. https://pubmed.ncbi.nlm.nih.gov/12117397/.

CHAPTER 3

Anahad O'Connor,"New York Attorney General Targets Supplements at Major Retailers," *New York Times,* February 3, 2015. https://archive.nytimes .com/well.blogs.nytimes.com/2015/02/03/new-york-attorney-general -targets-supplements-at-major-retailers/.

CHAPTER 4

Butow P., Price M., Coll J. et al. "Does stress increase risk of breast cancer? A 15-year prospective study." *Psycho-Oncology* (2018). https://onlinelibrary.wiley.com/doi/abs/10.1002/pon.4740#:~:text=Conclusions,methods%20of%20BCa%20risk%20reduction.

Schoemaker MJ, Jones ME, Wright LB et al. "Psychologic stress, adverse life events and breast cancer incidence: a cohort investigation in 106,000 women in the United Kingdom." *Breast cancer Research* (2016). https://pubmed.ncbi.nlm.nih.gov/27418063/.

CHAPTER 5

"Screening for Breast Cancer: Recommendations for the U.S. Preventive Services Task Force." *Annals of Internal Medicine* (2002). https://www.acpjournals.org/doi/10.7326/0003-4819-137-5_Part_1-200209030-00004;https://www.uspreventiveservicestaskforce.org/uspstf/recommendation/breast-cancer-screening-2002.

Susan Drossman, Elisa R. Port, Emily Sonnenblick, "Why the Annual Mammogram Matters," *New York Times*, October 28, 2015. https://www.nytimes.com/2015/10/29/opinion/why-the-annual-mammogram-matters.html.

"United States Preventive Service Task Force breast cancer screening recommendations 2009." https://www.uspreventiveservicestaskforce.org/uspstf/recommendation/breast-cancer-screening-2009.

"United States Preventive Service Task Force breast cancer screening recommendations final statement April 30, 2024." https://www.uspreventiveservicestaskforce.org/uspstf/recommendation/breast-cancer-screening#bcei-recommendation-title-area.

CHAPTER 6

"FDA Updates Mammography Regulations to Require Reporting of Breast Density Information and Enhance Facility Oversight," March 9, 2023. https://www.fda.gov/news-events/press-announcements/fda-updates-mammography-regulations-require-reporting-breast-density-information-and-enhance.

CHAPTER 7

Monticciolo D. L., Newell M. S., Moy L. et al. "Breast Cancer Screening for Women at Higher-Than-Average Risk: Updated Recommendations From the ACR." *Journal of the American College of Radiology* (2023). https://www.jacr.org/issue/S1546-1440(23)X0009-X.

CHAPTER 9

"US Food and Drug Administration," Breast Cancer Screening: Thermogram No Substitute for Mammogram. https://www.fda.gov/consumers/consumer-updates/breast-cancer-screening-thermogram-no-substitute-mammogram.

CHAPTER 10

"American Cancer Society," American Cancer Society Recommendations for the Early Detection of Breast Cancer" https://www.cancer.org/cancer/types/breast-cancer/screening-tests-and-early-detection/american-cancer-society-recommendations-for-the-early-detection-of-breast-cancer.html#:~:text=While%20the%20American%20Cancer%20Society,average%20risk%20of%20breast%20cancer.

CHAPTER 14

"National Cancer Institute," Breast Cancer Risk Assessment Tool: Online Calculator (The Gail Model). https://bcrisktool.cancer.gov/

"Ikonopedia," Breast Cancer Risk Calculator (IBIS). https://ibis.ikonopedia.com/.

CHAPTER 21

Gentilini O. D., Botteri E., Sangalli C. et al. "Sentinel Lymph Node Biopsy vs No Axillary Surgery in Patients With Small Breast Cancer and Negative Results on Ultrasonography of Axillary Lymph Nodes: The SOUND Randomized Clinical Trial." *JAMA Oncology* (2023). https://pubmed.ncbi.nlm.nih.gov/37733364/.

Giuliano A. E., Ballman K. V. McCall L. et al. "Effect of Axillary Dissection vs No Axillary Dissection on 10-Year Overall Survival Among Women With Invasive Breast Cancer and Sentinel Node Metastasis: The ACOSOG Z0011 (Alliance) Randomized Clinial Trial." *Jama* (2017). https://pubmed.ncbi.nlm.nih.gov/28898379/.

CHAPTER 23

"COMET study," COMET- a study for low risk DCIS: expanding knowledge and options. https://www.cometstudy.org/.

CHAPTER 24

Kalinsky K., Barlow W. E., Gralow J. R. et al. "21-Gene Assay to Inform Chemotherapy Benefit in Node-Positive Breast Cancer." *New England Journal of Medicine* https://pubmed.ncbi.nlm.nih.gov/34914339/.

Paik S., Shak S., Tang G. et al. "A multigene assay to predict recurrence of tamoxifen-treated, node-negative breast cancer." *New England Journal of Medicine* (2004). https://pubmed.ncbi.nlm.nih.gov/15591335/.

Sparano JA, Gray RJ, Makower DF et al. "Adjuvant Chemotherapy Guided by a 21-Gene Expression Assay in Breast Cancer." *New England Journal of Medicine* (2018). https://pubmed.ncbi.nlm.nih.gov/29860917/.

CHAPTER 26

Hughes KS, Schnaper LA, Bellon JR et al. "Lumpectomy plus tamoxifen with or without irradiation in women age 70 years or older with early breast cancer: long-term follow-up of CALGB 9343." *Journal of Clinical Oncology* (2013). https://pubmed.ncbi.nlm.nih.gov/23690420/.

CHAPTER 28

Schmid P, Cortes J, Dent R et al. "Overall Survival with Pembrolizumab in Early-Stage Triple-Negative Breast Cancer." *New England Journal of Medicine* (2024). https://pubmed.ncbi.nlm.nih.gov/39282906/.

CHAPTER 32

Partridge AH, Nima SM, Ruggeri M et al. "Interrupting Endocrine Therapy to Attempt Pregnancy after Breast Cancer." *New England Journal of Medicine* (2023). https://pubmed.ncbi.nlm.nih.gov/37133584/.

CHAPTER 33

Zaveri S, Nevid D, Ru M et al. "Racial Disparities in Time to Treatment Persist in the Setting of a Comprehensive Breast Center." *Annals of Surgical Oncology* (2022). https://pubmed.ncbi.nlm.nih.gov/35697955/.

INDEX

accelerated partial breast irradiation (APBI), 287–88

adjuvant therapy, 277

adoption, 134, 139, 142, 146

adriamycin ("red devil"), 262, 269, 306, 348

aesthetics, 192–93, 197, 206, 207, 211

aesthetic flat closure (AFC), 228

alcohol intake, 7–8, 32–33, 63

American Cancer Society (ACS), 71, 113, 114

anastrozole (Arimidex), 297, 303–4

anticancer diet, 23, 27–28

antihormonal treatment, 295–304, 319
  Amanda's story, 297–98
  aromatase inhibitors, 303–4
  cancer type and, 266, 276
  DCIS and, 242–43, 248, 282–83
  for high-risk patients, 159–61, 162–63
  HRT therapy and, 42
  lymph nodes and, 235
  for male breast cancer, 357–60, 364
  Mary Joe's story, 295
  Molly's story, 296–97
  potential benefits of, 298–99
  SERDs, 302

SERMs, 299–302. *See also* tamoxifen
  targeted therapies and, 312
  for young women, 343, 344–45
  young women and fertility preservation, 343, 344–45

aromatase inhibitors, 303–4

artificial intelligence (AI), 77, 92, 159, 304

Ashkenazi Jewish descent, 127, 128, 134, 140, 142

atypical ductal hyperplasia (ADH), 147–48

atypical hyperplasia, 147–48, 151, 154, 158–59, 161, 162

atypical lobular hyperplasia (ALH), 147–48

axillary dissections, 231–37, 279, 334, 363

axillary nodes, 212, 231

back pain, 206, 209

basics of breast cancer, 169–77
  cancer subtypes, 171–74
  stages, 175–76

benign lumps, 14, 16, 95, 96–99, 146

Bentley, Sarrah Strimel, 201–2, 206–7, 251–52, 258–59, 323, 342–44, 386

bilateral mastectomies. *See* double mastectomies

bioidentical hormones, 42–43
biopsies, 93–99
  atypical hyperplasia, 147–48, 151,
    154, 161, 162
  benign results, 16, 95, 96–99, 101,
    146
  Juana's story, 240–41
  Katherine's story, 111–12
  liquid, 105–7
  localization, 95–96
  Marlene's story, 144
  occult breast cancer, 333
  sentinel node, 233–37, 238, 245–46, 363
  Suzanne's story, 93–94, 97
  young women, 336–37
birth control, 39, 44–45, 152, 160,
    340–41
"black box" labeling, 43–44
Black women, 10, 74, 152, 273–74,
    347–55, 386–87. *See also* disparities
  Chiquita's story, 347–51, 386–87
blood clots, 41, 42, 160, 300, 301, 312
body scans, 103–5
bone density, 41, 55
brain fog, 39, 301
BRCA mutation, 128–34, 266
  Chris's story, 125–28, 132, 133
  high-risk situations, 157–58, 161
  Jill's story, 135–38
  mastectomy considerations, 197–98,
    204, 205, 206, 210
  PARP inhibitors, 311
  paternal side, 135–42
  Sam's story, 7–8
  Stacey's story, 305
  young women, 339
BRCA1 mutation, 128–33, 139–41,
    162, 361
BRCA2 mutation, 128–31, 133, 361
breast arterial calcifications (BAC), 78
Breast Cancer Awareness Month, 360
breast conservation strategy. *See*
    lumpectomies

breast density, 79–83, 121, 150–51
  MRIs and, 88–89, 150–51
breastfeeding
  after breast cancer, 345–46
  breast cancer risk and, 47, 152
  diet and Cindy's story, 22
  mastitis, 329
  nipple discharge, 14
  surgery considerations, 191
"breast friends," 178–84
  not having personal experience,
    182–83
  sharing your personal experience,
    180–82
breast health
  lifestyle factors, 9, 21–36
  overview of, 9, 13–17
breast implant-associated anaplastic
    large cell lymphoma (BIAALCL),
    225–26
breast implants, 222–27
breast pain, 115–16
breast reconstruction. *See*
    reconstruction
breast self-exams, 113–16, 121
  Katherine's story, 110–13
  recommendations, 113–15
breast skin, changes in, as warning
    sign, 15
caffeine, 54, 116
calcifications, 78, 87, 144, 240, 241, 247
calcium, 55
Capello, Nancy, 82
cardiovascular health. *See* heart health
Carter, Chiquita, 347–51, 386–87
cataracts, 300
CDK4/6 inhibitors, 312
cell cycle inhibitors, 312
chemoprevention, 159–61. *See also*
    antihormonal treatment
chemotherapy, 257–72
  cycles, 265
  de-escalation, 75, 265–66

chemotherapy (*cont.*)
   disparities and, 348–49
   Emma's story, 323–24
   Gina's story, 274–75, 321
   HER2/neu positive, 279–80
   immunotherapy and, 310–11
   Jill's story, 262, 269
   Katherine's story, 321–22
   male breast cancer, 357–58
   navigating "life" during, 4–5
   oncotype scores, 266–68
   Patricia's story, 273–74
   Sarrah's story, 258–59, 342–43
   side effects, 258, 259, 260, 262, 265,
      269–70
   Stacey's story, 257–58, 275
   surgery first or, 273–80
   triple negative, 268, 273, 274, 276,
      277–78
   Viviana's story, 259–61
   young women, 342–43, 344
chest wall radiation, 149–50, 157,
   292–93
childbearing, 151–52, 340–41, 342–46
children, talking with, 366–77
   about diagnosis, 370–72
   ages five and under, 372–73
   ages five to eleven, 373–74
   ages twelve to seventeen, 374–76
circulating tumor cells (CTCs), 105–7
circulating tumor DNA (ctDNA),
   105–7
clinical trials
   antihormonal medications, 159–61
   chemotherapy, 273–74, 275
   DCIS and watchful waiting, 250
   lymph node removal, 238
   male breast cancer patients, 364
   radiation, 286–87, 288–89
   supplements, 51, 53–54
cold capping, 262, 271–72, 377
colon cancer, 26, 61, 106, 107, 140,
   202–3, 341

communication, 366–80
   age eighteen and over, 376–77
   with children. *See* children, talking
      with
   with friends and family, 377–79
   work colleagues, 379–80
confocal microscopy, 103
core needle biopsies, 94
cortisol, 29, 60
COVID-19 pandemic, 57–58, 69,
   351–52, 356
Cures Act, 116–19
dairy milk, 341
daughters, talking with, 375–76
DCIS (ductal carcinoma in situ), 171,
   174, 175, 240–46
   Ariel's story, 247–48
   diagnosis, 240–42
   Hannah's story, 281–82, 289–90
   Juana's story, 240–41
   margins, 242, 243, 244, 290
   role of checking lymph nodes in,
      245–46
   treatment, 242–44
   watchful waiting, 248, 249–50, 282
   young women, 336
de-escalation, 75, 170, 239, 250,
   265–66, 286, 288, 289, 331
dense breasts. *See* breast density
diabetes, 26, 35, 351
diagnosis, 9–10. *See also* screening
   average age of, 191
   "breast friends," 178–84
   disparities and, 347–48, 352–53
   first consultation, 177
   male breast cancer, 362–63
   MRIs after, 89–90
   stages, 175–76
   subtypes and classifications, 171–74
   talking to children after, 370–72
DIEP flap surgery, 221–22, 227
diet, 22–23, 24, 26–27, 63
disparities, 10, 347–55

Chiquita's story, 347–51
cure and survival rates, 351–52, 354–55
diagnosis, 347–48, 352–54
double lumpectomies, 197, 244
double (bilateral) mastectomies, 200–209, 317
BRCA mutations, 131–32, 198
breastfeeding after, 345–46
Joanna's story, 202–3
Mary Joe's story, 200–201
reconstruction considerations, 221
regrets about, 181
Sarrah's story, 201–2, 206–7
drains, 215, 218
ductal carcinoma in situ. *See* DCIS
dysgeusia, 308
endocrine disruptors, 341
endocrine therapy. *See* antihormonal treatment
endogenous hormones, 41
energy level, 3, 39, 40, 284–85
environment. *See* lifestyle factors
Eskovitz, Katherine, 110–13, 321–22, 387–88
estrogen
antihormonal treatment, 298–99
exercise and, 29
therapy. *See* hormone replacement therapy
weight and, 25–26, 39
estrogen receptor (ER), 172–73, 242
Evert, Chris, 125–28, 132, 133
exercise, 21–22, 28–32, 63
chemotherapy and, 259–61
outdoor, benefits of, 30–32
false-positive test results, 72, 76, 78, 80, 88, 105, 106, 108, 121
family. *See also* "breast friends"
talking with, 377–79
family history (genetics), 128–34, 145–46
Chris's story, 125–28, 132

Christine's story, 153–55
disparities and, 347–48
Jenna's story, 143
male breast cancer and, 361
Marlene's story, 144–45
mastectomy considerations, 197–98, 206–7
paternal transmission, 135–42
risk factors and screening, 13, 15–16, 131–32, 145–46, 206–7. *See also* genetic testing
Sophie's story, 138–39
young women and, 336, 339
fatigue, 270, 284–85, 301, 311, 312
FDA, 43–44, 77, 83, 108, 159, 225, 226
FDG-PET scans, 27–28
feminine identity, 190–91, 192, 217, 219
Fernández, Mary Joe, 200–201, 295, 387
fertility preservation, 342–46
fibroadenoma, 93–94, 97, 99
fibrocystic disease, 114–16
fibroglandular breast tissue, 80–81
Figueroa, Viviana, 259–61, 313–14, 316–17, 389–90
FISH (Fluorescence In Situ Hybridization) test, 173
flap reconstruction, 221–22
"flat after mastectomy," 219, 220, 227–28
follow-up care, 313–20
three-month appointment, 314–16
friends. *See also* "breast friends"
talking with, 377–79
full-body scans, 103–5
Gail Model, 155–56, 158
Galleri test, 105–6
genetic predisposition. *See* BRCA mutation; family history
genetic testing, 9, 125–36, 163, 360
Chris's story, 125–28, 132, 133
Christine's story, 153–55
guidelines, 132–34

genetic testing (*cont.*)
  Hannah's story, 282
  Sam's story, 7–8
  Sarrah's story, 201–2
genomic profile tests, 266–68
ginkgo biloba, 51
GLP-1 receptor agonists, 34–36
Gormley, Emese, 109
gratitude, 387–88
green exercise, 30–32
Griffith, Stacey, 21–22, 69, 178–79, 275, 380
gynecomastia, 361–62
hair loss, 262, 270, 284, 322
  cold capping, 262, 271–72, 377
headaches, 39, 272, 301
health insurance. *See* insurance
heart disease, 26, 33, 35, 44, 285, 351
heart health
  exercise for, 29
  mammograms for, 77–78
heart, radiation and damage to, 285, 293
HER2/neu negative, 172–73, 174, 242, 312
  lymph node surgery, 237–38
  treatment considerations, 266, 268, 274, 276–77, 288
HER2/neu positive, 172–73, 174
  lymph node surgery, 237–38
  treatment considerations, 268, 276–77, 279–80, 291–92
high risk, 125–63
  genetic testing, 125–36
  MRIs for, 88–89
  screening situations, 153–59
Hodgkin's lymphoma, 149–50, 292–93
hormones, 60, 151–52
  weight and, 25–26, 39, 41
hormone-blocker medications. *See* antihormonal treatment
hormone replacement therapy (HRT), 37–45, 152

hot flashes, 37, 39, 43, 160, 297, 301, 303, 364
hypofractionation, 287
IBIS model, 155–56, 158
immune system, 29, 40, 56, 61
immunotherapy, 276, 310–11
implants, 222–27
impostor syndrome, 322–23
inflammatory breast cancer, 172, 174, 329, 330–31
insulin-like growth factor (IGF), 60
insurance
  disparities, 347, 353–54
  genetic testing, 132–33
  IVF, 344
  reconstruction, 220
  thermography, 108
  ultrasounds, 82–83
intraoperative radiation (IORT), 288
invasive ductal carcinoma, 171–74, 229–30, 333
invasive lobular carcinoma, 171–74, 333
in vitro fertilization (IVF), 134, 342, 344
joint pain, 29, 51, 303
joint replacements, 91
Jolie, Angelina, 132
"joy mining," 5, 386
Katz, Amanda, 217–19, 220, 297–98, 322–23
Ki-67 level, 173–74, 294
Klinefelter's syndrome, 361
lifestyle factors, 9, 21–36, 63
  alcohol, 32–33
  exercise, lack of, 28–32
  stress, 57–62
  weight, 25–28
  young women, 339–41
liquid biopsies, 105–7
lobular carcinoma in situ (LCIS), 148–49, 154, 158–59
localization, 95–96

"local recurrence," 195

lumpectomies, 194–97
  advantages of, 194–97
  DCIS, 240–41, 242–43
  decision-making considerations,
    190–94, 253
  follow-up care, 315, 317
  Hannah's story, 282
  Jennifer's story, 189–90
  Juana's story, 240–41
  margins, 195–97
  Paget's disease, 332–33
  recurrence risk, 90, 195–96, 197,
    243, 317–18
  survival rates, 194–95
  young women, 336

lung cancer, 61, 102–3, 105, 270, 348

lymphedema, 212, 232–33, 236, 238,
    274, 278, 292, 349, 350

lymph nodes, 230–31
  checking in DCIS, 245–46
  enlarged under the arm, as warning
    sign, 14–15, 330
  status and staging, 231, 233–34, 238

lymph node surgery, 229–39
  major advancements in, 231–37
  Marianne's story, 229–30, 236

lymphovascular invasion (LVI),
    173–74, 294

Lynparza, 306–9, 311–12

magnetic resonance imaging (MRIs),
    85–92, 170
  biopsies after, 94–95
  BRCA mutations, 131–32
  breast density, 88–89, 121, 150–51
  high-risk situations, 157–58
  long-term follow-up, 90
  Meredith's story, 85–86
  occult breast cancer, 333–34
  recommended scenarios for, 87–90
  when not to get, 91–92

male breast cancer, 356–65, 388–89
  diagnosis, 362–63
  genetic predisposition, 129, 131
  incidence, 360–61
  risk factors, 361–62
  surgery, 363–64
  treatment, 364–65

MammaPrint, 266–68

mammograms, 67–78
  age for, 70–71
  benefits of yearly, 68, 70, 71–75, 121,
    156
  biopsies after, 94–95
  BRCA mutations, 131–32
  breast density and, 79–83
  contrast-enhanced, 76–77, 103,
    169–70
  DCIS, 240–42
  high-risk situations, 156
  Jennifer story, 67–69
  Juana's story, 240–41
  Katherine's story, 110–12
  Marlene's story, 144–45
  MRIs vs., 86–87, 88
  occult breast cancer, 333
  Stacey's story, 69
  3D, 76, 103, 169–70

mantle radiation. See chest wall
  radiation

margin status, 195–97, 216, 242, 243,
    244, 290, 383

Margolin, Evan, 356–58

marijuana, 362

Martin, Jill, 135–238, 262, 269

mass or lumps, as warning sign, 13–14,
    97

mastectomies, 197–99, 210–15
  advantages of, 197–98
  Alison's story, 187–89
  breastfeeding and, 345–46
  for DCIS, 243–44, 245–46
  decision-making considerations,
    190–94, 204–8, 253
  drains, 215, 218
  follow-up care, 315, 317

mastectomies (*cont.*)
genetics and BRCA mutations, 131–32, 197–98
for inflammatory breast cancer, 331
Jennifer's story, 189–90
Jess's story, 6
Joanna's story, 202–3
for male breast cancer, 363–64
Melanie's story, 210–11
modified radical, 212, 213, 331
nipple-sparing, 210–11, 214–15
Paget's disease, 332–33
post-mastectomy radiation treatment (PMRT), 292
prophylactic (risk-reducing), 131–32, 161, 163
reconstruction considerations, 211–15, 216–17, 219–28
regrets about, 181
Sam's story, 7–8
Sarrah's story, 201–2, 206–7
single vs. double, 200–209
skin-sparing, 213–14
Susan's story, 1–3
types of, 211–15
mastitis, 329
Medicaid, 349–50, 353–54
medications. See also specific medications
exercise and, 29–30
medullary carcinoma, 172, 174, 334
menopause
HRT and, 37–38, 41–42
symptoms, 37, 39–44, 160, 303
weight gain and, 33–34
menstruation (menarche), average age to begin, 46, 151, 340
mental health, 297, 313–14
exercise for, 29, 31
metastatic disease. *See* stage IV
milk ducts, 47, 98–99, 171, 244
misinformation, 6–8, 24, 49–50, 95, 284, 304

modified radical mastectomies, 212, 213, 331
MRIs. *See* magnetic resonance imaging
mucinous carcinoma, 172, 174, 334
myocarditis, 311
nausea, 270, 301, 308, 309, 311, 312, 388
needle biopsies, 94–95, 97, 144
neoadjuvant therapy, 277–79
nerve grafts, 215
night sweats, 37, 39, 40, 297, 301, 303, 364
nipple discharge, 14, 98–99, 332
nipples, changes in, as warning sign, 15
nipple-sparing mastectomies, 210–11, 214–15
nipple tattooing, 213, 218, 252, 357, 363
noninvasive breast cancer. *See* DCIS
nulliparity, 151–52
Obama, Barack, 117
obesity (overweight), 25–28, 33–36, 339–40, 362
occult breast cancer, 333–34
Oncotype DX, 266–68, 276, 295, 297, 357, 359–60, 364
outdoor exercise, 30–32
ovarian cancer, 43, 45, 107, 270, 304, 336, 367
genetics and family history, 125–30, 132, 133–34, 139–41, 142
novel therapies, 305–6, 307
overdiagnosis, 78, 90
overtreatment, 78, 247–50, 264
Paget's disease, 332–33
pain medications, 8, 189, 208, 379
pancreatic cancer, 100–102, 129–30, 134, 140, 270
papillomas, 98–99, 147–48
"parentification," 375
PARP inhibitors, 311–12
partial mastectomies. *See* lumpectomies

PASH (pseudoangiomatous stromal hyperplasia), 97–98
paternal transmission, 135–42
peau d'orange, 330
pembrolizumab (Keytruda), 310–11
PET scans, 27–28, 176, 182, 279–80, 353
phyllodes tumors, 99
phytoestrogens, 42–43
placebo effect, 43, 51, 52, 56
plastic surgeons, 211, 217, 220, 227–28
    implants, 222, 223–27
    Sarrah's story, 202
plastic surgery. *See* reconstruction
pneumonitis, 311, 348
post-mastectomy radiation treatment (PMRT), 292
pregnancy. *See also* childbearing
    fibroadenomas and, 97
premenopausal women. *See* young women
probiotics, 51–52
progesterone receptor (PR), 172–73, 242, 268
prophylactic mastectomies, 131–32, 161, 163
prostate cancer, 107, 129, 130, 134, 139, 248–49
PSA (prostate-specific antigen) tests, 107
radial scars, 98
radiation therapy, 281–94
    APBI, 287–88
    Black women and, 350
    for DCIS, 243
    disparities and, 348–49, 350
    follow-up, 316, 319
    Hannah's story, 281–82, 281–83, 289–90
    for inflammatory breast cancer, 331
    lumpectomies and, 195
    not advisable, 149–50, 292–94
    omitting, 288–90

recurrence risk, 286, 288–92
    Sarrah's story, 343
    shorter courses, 286–87
    side effects, 283–86
    for young women, 343, 344
raloxifene, 160–61
reconstruction, 216–28
    after inflammatory breast cancer, 331
    after lumpectomy, 196–97
    after male breast cancer, 363
    Amanda's story, 217–19, 220
    flap, 221–22
    Hilary's story, 216–17, 220, 222
    implants, 222–27
    mastectomy type and, 211–15, 216–17, 219–20
    radiation and, 286
    Sarrah's story, 251–52
recurrence risk, 317–20
    antihormonal therapy and, 298, 299, 300, 319, 344–45
    chemotherapy and, 269, 271
    lifestyle factors and, 25–27, 28, 32, 34, 36, 49, 59
    lumpectomy and, 90, 195–96, 197, 243, 317–18
    lymph node surgery and, 230, 235, 236
    mastectomies and, 197–98, 204, 205, 207, 210, 224, 244
    radiation and, 286, 288–92
    tumor profile tests, 266–68
regrets, 181, 249
risk factors, 3–4, 8, 143–52
    family history. *See* family history
    high risk. *See* high risk
    hormone-based, 45–47
    lifestyle. *See* lifestyle factors
    male breast cancer, 361–62
    young women, 339–41
risk models, 154–56, 158–59
Runfola, Gina, 366–67, 381–85

Sager, Stacey, 162, 257–58, 305–9

saline implants, 223, 225

Schneiderman, Eric, 52–53

sclerosing adenosis, 98

screening, 9, 15–17, 67–122

  biopsies. *See* biopsies

  BRCA mutations, 131–32

  cutting-edge technology, 100–109

  disparities and, 349–50

  high-risk situations, 153–59

  inflammatory breast cancer, 330

  MRIs. *See* magnetic resonance
    imaging

  occult breast cancer, 333–34

  Paget's disease, 332

  risk models, 154–56, 158–59

  timing, 121–22

  ultrasounds. *See* ultrasounds

"secondary malignancy," 285

second opinions, 188

selective estrogen receptor degraders
  (SERDs), 302

selective estrogen receptor modulators
  (SERMs), 299–302. *See also*
  tamoxifen

self-exams. *See* breast self-exams

self-image, 190–91, 192, 217, 219

sentinel node biopsies, 233–37, 238,
  245–46, 363

sex life, 39, 191–92, 364

silicone implants, 225–26

simple (total) mastectomies, 212–13

Sims, Molly, 109

skin dimpling or changes, as warning
  sign, 14

skin-sparing mastectomies, 213–14

sleep disturbances, 3, 16, 37–38, 40–41,
  58

smoking, 60, 79, 102, 105, 153, 362

sonograms. *See* ultrasounds

Sparano, Joseph, 75

stages, 175–76, 231

stage IV, 125, 127, 175, 176, 263, 318

stereotactic biopsies, 94–95

Storm, Hannah, 120, 253, 281–83,
  289–90

stress, 57–62

  Andrea story, 57–59, 61–62

  exercise for, 30–31

  research studies, 59–61

subtypes, 171–74

sugar, 22–23, 24, 27–28

sunlight, 31, 55

superfoods, 27

supplements, 24, 48–56

  purported benefits of, 50–52

  quality factor, 52–53

  quantity factor, 53–54

  safety issues, 54

  side effects, 52

  Terri story, 48–49, 50–51, 55–56

support network, 5–6, 8–9. *See also*
  "breast friends"

  communication. *See*
  communication

surgeons. *See also* plastic surgeons

  decision-making considerations,
  190–94

  first consultation, 177

surgery, 187–215

  Alison's story, 187–88

  chemotherapy first or, 273–80

  follow-up care, 313–20

  inflammatory breast cancer, 331

  Jennifer's story, 189–90

  lumpectomy vs. mastectomy,
  187–99

  lymph node, 229–39

  male breast cancer, 359–60, 363–64

  Mary Joe's story, 200–201

  Patricia's story, 273–74

  preemptive, 161

  Sarrah's story, 201–2, 206–7

surrogacy, 345

survival rates, 4, 171

  disparities and, 351–52, 354–55

lumpectomies, 194–95
  mammograms and, 74, 75
swoopers, 183–84
tamoxifen, 299–302. *See also*
    antihormonal treatment
  for high-risk patients, 160–61,
    162–63
  for male breast cancer, 357–60
  side effects, 300–301
  young women and fertility
    preservation, 343, 344–45
targeted therapies, 311–12
targeted ultrasounds, 84, 236
teenagers, talking with, 374–76
textured implants, 225–26
thermography, 107–9
thyroid cancer, 284
thyroiditis, 311
tissue expanders, 223–24
trans patients, 365
treatment, 170–71
  antihormonal. *See* antihormonal
    treatment
  chemotherapy. *See* chemotherapy
  DCIS, 242–44
  disparities and, 349–50
  follow-up care, 313–20
  inflammatory breast cancer, 331
  male breast cancer, 364–65
  novel therapies, 305–12
  overtreatment, 247–50, 264
  radiation. *See* radiation therapy
  surgery. *See* surgery
triple negative, 130–31, 173, 174,
    277–78
  Black women, 355
  chemotherapy, 268, 273, 274, 277–78
  lymph node surgery, 238
  treatment considerations, 268, 273,
    274, 276, 291–92
tubular carcinoma, 172, 174, 334
tummy tucks, 217
tumor profiling, 266–68

21st Century Cures Act, 116–19
Tyrer-Cuzick model, 155–56, 158
ultrasounds, 79–84, 170
  biopsies after, 94–95, 97
  breast density and, 121, 150–51
  high-risk situations, 157–58
  Katherine's story, 110–12
  occult breast cancer, 333
  Robin's story, 83–84
  Vanessa's story, 79–80, 82
United States Prevention Service Task
    Force (USPSTF), 71, 74, 113–14,
    355
uterine cancer, 26, 41, 160, 284, 300
vaginal dryness, 39, 303
vegetarian (vegan) diet, 48, 49, 63
vitamins. *See* also supplements
  deficiencies, 54–55
  vitamin B12, 55
  vitamin C, 55
  vitamin D, 31, 55
warning signs, 13–15
watchful waiting, 78, 149, 247–50, 282
weight control/loss, 26–27, 33–36, 63
weight factors, 25–28
weight gain, 39, 60, 265, 270, 301, 357
  chemotherapy and, 258–59, 324
Women's Health and Cancer Rights
    Act (WHCRA), 220
work colleagues, talking with, 379–80
young women, 303, 335–41
  antihormonal treatment, 300, 303–4
  fertility preservation, 342–46
  Gina's story, 335–37
  incidence, 339
  Melissa's story, 337–39
  risk factors, 339–41
  Sarrah's story, 342–43
zinc, 55